HYDROLOGIC
SCIENCES

Taking Stock and Looking Ahead

Proceedings of the 1997 Abel Wolman Distinguished Lecture
and Symposium on the Hydrologic Sciences

Water Science and Technology Board

Commission on Geosciences, Environment, and Resources

National Research Council

NATIONAL ACADEMY PRESS
Washington, D.C.   1998

NATIONAL ACADEMY PRESS • 2101 Constitution Avenue, N.W. • Washington, DC 20418

NOTICE: The project that is the subject of this report was approved by the Governing Board of the National Research Council, whose members are drawn from the councils of the National Academy of Sciences, the National Academy of Engineering, and the Institute of Medicine. The members of the committee responsible for the report were chosen for their special competences and with regard for appropriate balance.

This report has been reviewed by a group other than the authors according to procedures approved by a Report Review Committee consisting of members of the National Academy of Sciences, the National Academy of Engineering, and the Institute of Medicine.

Support for this project was provided by the Department of Energy under Agreement No. DE-FG02-93ER61547, the National Science Foundation under Agreement No. BES-9812445, the U.S. Geological Survey under Agreement No. 98HQAG2023, and the National Water Research Institute. Any opinions, findings, conclusions, or recommendations expressed in this publication are those of the author(s) and do not necessarily reflect the view of the organizations or agencies that provided support for this project.

Library of Congress Catalog Card Number 98-84689
International Standard Book Number 0-309-06076-1

Additional copies of this report are available from:

National Academy Press
2101 Constitution Ave., NW
Box 285
Washington, DC 20055
800-624-6242
202-334-3313 (in the Washington metropolitan area)
http://www.nap.edu

THE ABEL WOLMAN DISTINGUISHED LECTURER

THOMAS DUNNE, University of California, Santa Barbara

SYMPOSIUM MODERATOR

DAVID L. FREYBURG, Stanford University

SYMPOSIUM CONTRIBUTORS

DIANE M. McKNIGHT, University of Colorado
ERIC F. WOOD, Princeton University
FRED M. PHILLIPS, New Mexico Institute of Mining and Technology
STEPHEN J. BURGES, University of Washington

SYMPOSIUM DISCUSSANTS

KAYE L. BRUBAKER, University of Maryland
DARA ENTEKHABI, Massachusetts Institute of Technology
DAVID P. GENEREUX, Florida International University
EFI FOUFOULA-GEORGIOU, University of Minnesota

SYMPOSIUM COORDINATORS

LAURA EHLERS, Program Officer
ANITA HALL, Production Assistant
SHEILA DAVID, Program Officer (through 6/97)

Acknowledgments

This report has been reviewed by individuals chosen for their diverse perspectives and technical expertise, in accordance with procedures approved by the NRC's Report Review Committee. This independent review provided candid and critical comments that assisted the authors and the NRC in making the published report as sound as possible and ensured that the report meets institutional standards for objectivity, evidence, and responsiveness to the study charge. The content of the review comments and draft manuscript remain confidential to protect the integrity of the deliberative process. We wish to thank the following individuals for their participation in the review of this report:

Raphael Bras, Massachusetts Institute of Technology
George Hornberger, University of Virginia
Marc Parlange, The Johns Hopkins University
Suresh Rao, University of Florida
M. Gordon Wolman, The Johns Hopkins University

While the individuals listed above have provided many constructive comments and suggestions, responsibility for the final content of this report rests solely with the authoring committee and the NRC.

Contents

Overview

In 1991, the Water Science and Technology Board's Committee on Opportunities in the Hydrologic Sciences (hereafter referred to as "the committee"), completed its report, *Opportunities in the Hydrologic Sciences* (National Academy Press, Washington, D.C.). Under the leadership of Peter S. Eagleson, a group of our colleagues crafted a report both sophisticated and engaging—a thoughtful reflection on hydrologic science as a geoscience; an articulation of a vision of research, education, and institutional support in the hydrologic sciences; and a rousing call to action. Over the past six years this report, sometimes known as the "Blue Book," has indeed stimulated actions and a great deal of discussion. It has resulted in the creation of a Hydrologic Sciences Program within the Earth Sciences Directorate of the National Science Foundation and has been cited as a key component in the founding of several hydrologic sciences research and teaching programs in U.S. universities.

At its heart, *Opportunities in the Hydrologic Sciences* articulates a vision of hydrologic science as a distinct geoscience with a crucial societal role. A key objective of the committee in writing the report was to delineate a "puzzle-driven" science underlying, and complementary to, the "problem-driven" application of that science and attendant technology to problems of water resources. Although the discipline does overlap and "interact" with ocean, atmospheric, and solid earth sciences, as well as ecosystem sciences, the committee views hydrologic science as distinct and unique among the sciences in its focus on continental water processes and the global water balance. This focus leads to a set of fundamental concepts that form a cornerstone of the science and that are central to understanding all of the diverse processes at work in the global water balance.

1

These ideas include the concepts of fluxes, reservoirs, pathways, and residence times, along with the inevitability of dealing with an extraordinary range of spatial and temporal scales.

In *Opportunities in the Hydrologic Sciences* this vision is developed, illustrated, and supplemented by presenting representative critical and emerging areas of hydrologic research; by explicitly considering the role of data collection; by considering the nature of an education in hydrologic science; and finally by analyzing priorities, resources, and strategies important to the development of this fledgling geoscience. The report's systematic discussion of research areas includes some that indeed were very active in 1991 and other key lines of research that at the time were in need of nurturing and development. Because hydrologic science is inherently an observational science, the committee emphasizes the central role of data collection, distribution, and analysis in all of the critical and emerging areas of research. The committee also makes a strong case that a distinct hydrologic science cannot exist without an appropriate educational infrastructure. The report presents an approach to hydrologic science education that instills a sense of the internally driven intellectual pursuit of scientific understanding for its own sake and for that of society. Hydrologic education must be multidisciplinary, with programs building on a solid base of education in mathematics and the physical and life sciences at the graduate, undergraduate, and K-12 levels. Finally, the report articulates a set of priority problems that dominate hydrologic science at present. It also makes clear that resources must be redeployed and reorganized if the committee's recommendations are to be achieved.

THE SYMPOSIUM

One recommendation was to review "in five years the progress toward achieving the goals elaborated in this report, assessing the vitality of the field, surveying the changes that have occurred, and making recommendations for further action" (p. 16). Beginning in 1995 the Water Science and Technology Board began to consider whether and how to implement this recommendation. The board concluded that a review of the scope envisioned in the committee's recommendation was somewhat premature, since many of the recommendations are just being implemented and the results of research inspired by the committee are still sparse. Further, it is clear that the substantive recommendations of the report have not become outdated. Instead, the board decided to use the occasion of the seventh Abel Wolman Distinguished Lecture to direct the hydrologic community's attention to the vitality of the hydrologic sciences. The board attempted to do this in two ways: first, by challenging the Wolman Lecturer to use the bully pulpit of the Wolman Lecture to reflect freely and personally on the intellectual vitality of the hydrologic sciences; and second, by organizing a one-day symposium that would build on the Wolman Lecture by expounding on some

Abel Wolman
(1892-1989)

In the course of an exceptionally long and active career, Abel Wolman may have done more than any single person to bring the benefits of hydrologic science to the people of the world.

Wolman was born in Baltimore, Maryland, on June 10, 1892, the fourth of six children of Polish immigrants. He received a Bachelor of Arts degree from The Johns Hopkins University in 1913 and hoped to become a physician. Instead, his family persuaded him to enroll in Johns Hopkins University's newly opened School of Engineering, where he received his Bachelor of Science in Engineering with the first graduating class in 1915. (Wolman later founded what is today the Department of Geography and Environmental Engineering at Johns Hopkins and was a long time professor there.) His professional career had already started; he had begun collecting water samples on the Potomac River for the U.S. Public Health Service in 1912. After a year he joined the Maryland Department of Health, beginning an association that lasted until 1939.

During his period of state employment he performed some of his most distinguished scientific research in water purification. When Wolman began this work, his own family practiced water purification by tying a piece of cheesecloth around the spigot in their home to filter out stones and dirt that flowed through the city's water supply. Not only was the quality of drinking water in general highly variable and questionable, but water supply sources and waste disposal sites were also frequently the same. Outbreaks of waterborne diseases struck Baltimore and other American cities with alarming regularity.

Wolman worked with chemist Linn H. Enslow in the Maryland Department of Health to perfect a method of purifying water with chlorine at filtration plants. Although the idea of using chlorine as a purifying agent was not new, procedures were crude and produced wildly fluctuating water products. Wolman and Enslow developed a chemical technique for determining how much chlorine should be mixed with any given source of water, taking into consideration bacterial content, acidity, and other factors related to taste and purity. That collaboration produced the gift of safe drinking water for millions of people around the world.

Wolman's capacity and enthusiasm carried him into national and international service for a period that spanned six decades. A member of the first delegation to the World Health Organization (WHO), he worked on water supply, wastewater, and water resources problems throughout the world with WHO and the Pan American Health Organization. A consultant to Sri Lanka (then Ceylon), Brazil, Ghana, India, and Taiwan, he also chaired the committee that planned the water system for the new state of Israel, and he helped Latin American nations develop ways to finance their water systems. Abel Wolman was truly a man who transcended political and social boundaries and made the world a more livable place.

of the current efforts in the hydrologic sciences in celebration of the sixth anniversary of the Blue Book.

It seems particularly apropos to use the Wolman Lecture for this purpose. Abel Wolman (1892-1989) is the subject of the very first biographical vignette included in *Opportunities in the Hydrologic Sciences*. The vignette begins by stating that "in the course of an exceptionally long and active career, Abel Wolman may have done more than any single person to bring the benefits of hydrologic science to the people of the world." Dr. Wolman blended a sophisticated knowledge and curiosity about hydrologic and environmental sciences with an abiding concern for the institutions necessary to secure adequate and safe supplies of water for society. (The complete vignette from *Opportunities in the Hydrologic Sciences* is included here as a sidebar on p. 3) It seems an appropriate tribute to a great scientist and a stimulating and sophisticated report to bring them together for a day of reflection and review.

This, then, was the starting point for the 1997 Wolman Lecture and Symposium on the Hydrologic Sciences and the papers contained in this volume. These papers both explicitly and implicitly assess progress on a number of issues from the perspective of particular "puzzle areas" in hydrologic science. Some issues cross-cut all of these puzzle areas and so receive attention from all of the authors, whereas other issues are more focused and are addressed by a subset of the authors.

The board was honored that Thomas Dunne of the School of Environmental Science and Management and the Department of Geological Sciences and Geography at the University of California, Santa Barbara, agreed to present the 1997 Abel Wolman Distinguished Lecture. There are few hydrologists at work today who have not been influenced in one way or another by Dunne's thinking and writing. A member of the National Research Council's Committee on Opportunities in the Hydrologic Sciences, Professor Dunne presented a fascinating and thoughtful critique, the text of which is included here.

The symposium featured four particularly active areas of hydrologic inquiry. These areas were identified as emerging and critical in *Opportunities in the Hydrologic Sciences* and also have been characterized by significant progress over the past decade. These topical areas, as identified by the titles of papers presented at the symposium, are: "Aquatic Ecosystems: Defined by Hydrology," by Diane M. McKnight; "Hydrologic Measurements and Observations: An Assessment of Needs," by Eric F. Wood; "Ground Water Dating and Isotope Chemistry," by Fred M. Phillips; and "Streamflow Prediction: Capabilities, Opportunities, and Challenges," by Stephen J. Burges. The goal in choosing these four topics and speakers was not, of course, to be comprehensive or to suggest that these topics are in any sense of highest priority, either in intellectual value or in centrality to hydrologic science. Rather, it was hoped that an assessment of these topics would illustrate the state of the hydrologic sciences half a decade after the publication of *Opportunities in the Hydrologic Sciences*. To add more depth and breadth to discussions about the vitality of the hydrologic sciences, four discussants were invited to reflect on these primary papers. For these discussants the board enlisted several younger scientists who have already had an impact on the hydrologic sciences and who will likely be intellectual leaders in the field many years into the future: Kaye Brubaker, Dara Entekhabi, David Genereux, and Efi Foufoula-Georgiou. These short informal reflections are not included here but they stimulated considerable discussion at the symposium and have informed this overview.

THE WOLMAN LECTURE

Professor Dunne in his lecture provides a fairly explicit assessment of progress toward achieving the Committee on Opportunities in the Hydrologic Sciences' vision of hydrology as a distinct geoscience. His perceptive analysis and eloquent discussion provides insight into the nature of hydrologic science, along with a framework from which to contemplate the contributions of the other four symposium speakers. Professor Dunne's analysis and vision seem thoroughly consistent with that of the committee.

Dunne makes a persuasive case that, while hydrology has recently become recognized as a key element of the Earth sciences, the organizational and research infrastructure, educational institutions, and available funding have not caught up with appreciation for the science. Drawing examples from the increasingly important fields of planetary-scale hydrology and landscape hydrology, Dunne identifies and assesses four key elements essential to the sustenance of hydrologic science:

- the need to construct a coherent body of transferable theory and create an intellectual center for the science;
- the need for communication across multiple disciplines, backgrounds, and approaches;
- the need for appropriate measurements and observations; and
- the need for some level of central guidance and assessment.

The proliferating diversity of approaches to hydrologic puzzles, as opposed to a convergence on a consistent set of theoretical constructs, indicates to Dunne that hydrologic science has not yet reached maturity as a geoscience. He finds further support for this conclusion in the continuing lack of communication, and even intellectual respect, across the artificial boundaries of experimentalist versus theorist, modeler versus field observer, engineer versus earth scientist. The promise and excitement of the growing number of new data sources available to hydrologists are tempered by the recognition that hydrologists have not yet found a way to participate fully in the design of data collection campaigns so that they may contribute directly to the advancement of fundamental hydrologic science. Finally, Dunne contemplates the value of oversight and an institutional focal point for the convergence and continuity that would mark hydrologic science as a distinct and mature geoscience.

Dunne also introduces a number of key themes that appear throughout the papers of the symposium speakers and the comments of the discussants. These include the central role of scaling in developing understanding across the vast array of spatial and temporal scales of interest to hydrologic science; the importance of basing hydrologic science on an understanding of fundamental physical, chemical, and biological processes; the interpenetration of hydrology with the other Earth and life sciences; and the dominance of climatological and meteorological uncertainties over process uncertainties, particularly at the larger spatial scales of importance in regional, continental, and global hydrology. The four symposium papers in this volume examine these and other issues from several different perspectives.

AQUATIC ECOSYSTEMS: DEFINED BY HYDROLOGY

Use of the hydrologic cycle as a useful framework for interpreting and understanding biological and ecological processes is a recurrent theme throughout *Opportunities in the Hydrologic Sciences.* One of the committee's Critical and Emerging Areas is entitled "Hydrology and Living Communities." Further, the role of hydrology as a framework for aquatic ecology runs through all five prioritized research areas in the report's chapter on scientific priorities. Ecological processes either play a key role along with hydrologic processes in the research area or the research area is motivated by the role of hydrology in defining aquatic ecosystems. It is clear that no assessment of the state of hydrologic science can ignore the important role of aquatic ecosystems and ecological processes.

Diane McKnight is emphatic in declaring the important role of hydrologic science as an underpinning of aquatic ecosystem science, titling her talk "Aquatic Ecosystems: Defined by Hydrology." She notes that fluxes of water, mass, energy, and organisms do indeed define aquatic ecosystems. Analysis and prediction of these fluxes play a dominant role in understanding aquatic ecosystems, and hydrologic processes control many of these rates. In addition, many important integrative concepts in ecosystem science derive from the topology and directionality of hydrologic systems. "Upstream" and "downstream" matter in aquatic ecology just as much as they matter in hydrology. Finally, although not emphasized by McKnight, it is equally clear that ecological processes play much more than a passive role in controlling hydrologic processes.

In assessing hydrologic science vis-à-vis aquatic ecosystems, McKnight emphasizes the complexities of spatial and temporal scales and of scale mismatches, in both space and time. For example, she notes that scale mismatches can lead to hydrologic data collection designs which miss or fail to resolve key ecological fluxes and processes. Also, microscopic processes can play key roles in macroscopic behavior, either hydrological or ecological, with all the attendant challenges of scaling and parameterization.

McKnight's consideration of the role of hydrologic science in ecosystem science raises another important issue, one that threads its way through all of the papers in this volume: the joy and difficulty of multidisciplinarity. The complexity and scope of the hydrologic sciences, along with ecosystem sciences, provide ample opportunity for contributions from many different disciplines. The joy rests in the synergism of multiple perspectives, approaches, and backgrounds. Difficulties arise, however, because educational institutions, research funding entities, professional societies, and other relevant institutions are often organized along more traditional disciplinary lines.

HYDROLOGIC MEASUREMENTS AND OBSERVATIONS: AN ASSESSMENT OF NEEDS

Eric Wood returns to and focuses on a theme introduced by Tom Dunne that underlies all of the symposium papers as well as the Blue Book—the role of observation and measurement as the foundation of hydrology. This topic was considered important enough to merit its own chapter in the Blue Book, despite its role in most of the other chapters. Similarly it was afforded a key role at the symposium, despite its appearance in all of the other papers.

In his paper Wood demonstrates the centrality of measurements and observations to hydrologic science, using examples drawn from surface water hydrology, while attempting to understand why observations and measurements have become the "stepchild" of the science. His assessment of the "data health" of hydrologic sciences carries several themes. Despite a broad and general recognition of the centrality of measurements and observations in hydrology, specific quality and quantity criteria still do not exist against which to evaluate data collection efforts. In Wood's paper are echoes of Dunne's concern over the lack of convergence on fundamental principles, which remains a problem in hydrologic science. Echoing McKnight, but in the context of land surface-atmosphere interactions, Wood highlights the mismatches that can occur between data collected for one science (atmospheric science in this case) and the needs of another (hydrologic science). Finally, he finds that collaboration between data collectors/ observers and theoreticians/model developers has improved but must continue to be nurtured.

Wood's conclusion is mixed. Large quantities of new data have become available in the last decade, but there is still no coherent program of hydrologic data collection driven by the needs of hydrologic science. As discussant Dara Entekhabi stated at the symposium, hydrology has "transitioned from an era characterized by data starvation to [one characterized by] data confusion."

GROUND WATER DATING AND ISOTOPE CHEMISTRY

Environmental tracers received relatively little direct attention in *Opportunities in the Hydrologic Sciences* but played an important role there nonetheless. Environmental isotopes were identified as a key tool in studying subsurface hydrology, and careful reading reveals that tracers are implicitly an important component of a majority of the Categories of Scientific Opportunity described in Chapter 6 of the Blue Book.

In his paper Fred Phillips demonstrates nicely how germane the development of environmental tracers, particularly in subsurface hydrology, has been to the development of hydrology as a unique earth science. Environmental tracers, with applications to the unsaturated subsurface, shallow aquifers, deep aquifers, and regional flow systems, as well as surface-subsurface interactions, have achieved new prominence as much of the focus of hydrologic applications has shifted from water supply to water quality. Once again, central issues are characterizing fluxes, reservoirs, and change, and the broad range of spatial and temporal scales over which the physico-chemical processes controlling tracer movement occur.

In assessing ground water dating and isotope chemistry as components of the hydrologic sciences, Phillips, like his fellow authors, finds that multidisciplinarity plays a key role. In this case, other sciences, especially chemistry and geology, may drive the development of instrumentation and methodology, but the application to hydrologic problems becomes a key element of progress in the hydrologic sciences, leading to important hydrologic insights.

Progress in the development and application of environmental tracers to hydrologic problems is driven by instrumentation, funding, curiosity, practical need, and, not insignificantly, individual circumstances. Phillips illustrates this by describing a technique with both applied and basic applications (CFC tracers) that withered for a protracted period of time after initial discovery and development and an idea that developed purely out of scientific curiosity (^{36}Cl in fossil rat urine) but that almost immediately became applied to an important societal problem. The state of the science clearly depends on the scientific infrastructure and the development and maintenance of laboratory and experimental arts. Long-term, instrumentation-intensive funding is required. Phillips concludes that environmental tracer research and application are flourishing, playing a unique role in the advancement of hydrologic science. At least part of the credit for that, according to Phillips, is due to *Opportunities in the Hydrologic Sciences*, where the employment of modern geochemical techniques to trace water pathways was noted as an important emerging opportunity.

STREAMFLOW PREDICTION: CAPABILITIES, OPPORTUNITIES, AND CHALLENGES

As has been demonstrated many times, successful hydrologic prediction does not require complete scientific understanding. However, a necessary criterion of scientific understanding is the ability to predict. Therefore, the words "prediction" and "forecasting" appear repeatedly throughout *Opportunities in the Hydrologic Sciences*. Certainly runoff and streamflow prediction remains central to hydrologic science, fundamental as it is to, for example, geomorphic processes and landscape evolution, local and regional water balances, biogeochemical cycling, and ecological systems. Furthermore, many of the applications that drive public interest in and support the hydrologic sciences rest on the ability to predict streamflow.

In his paper Stephen Burges tackles the scientific and pragmatic challenge of streamflow prediction. While clearly articulating the societal need and hydrologic understanding necessary for successful streamflow prediction, Burges identifies and echoes a number of recurrent themes. Once again, we learn of the central role of the huge range of relevant spatial and temporal scales over which hydrologic processes act. Although much progress has been made in understanding how to apply knowledge across varying embedded scales, we are far from having an understanding that is adequate to allow successful streamflow prediction. Returning to a point noted by both Dunne and Wood, Burges emphasizes the importance of precipitation prediction to streamflow prediction. He suggests that understanding precipitation is the weak link in the problem of runoff prediction or, at the very least, that our understanding of terrestrial hydrologic processes cannot be truly tested predictively until precipitation forecasting uncertainty no longer dominates streamflow prediction uncertainty. Burges also reemphasizes the essential nature of observation and data. For streamflow prediction, precipitation data become central. Finally, streamflow prediction provides yet another example of the multidomain nature of hydrologic science. The atmosphere, oceans, hydrosphere, and biosphere all play important roles in continental water processes and the global water balance. Successful streamflow prediction is impossible without knowledge from all of these domains, and the "grand challenge" of hydrologic science is the coherent coupling of knowledge in all of these domains across a full range of spatial and temporal scales.

SUMMARY

Taken together, the Wolman Lecturer and the symposium speakers and discussants provide a consistent diagnosis of the vitality of the hydrologic sciences; the science is indeed a vital, intellectually challenging geoscience. However, it remains a young science, in need of greater coherence and struggling to cope with its multidisciplinary, multidomain nature. Diligence and vigilance in nurturing our science are essential. The vision of the Committee on Opportunities in the Hydrologic Sciences has stood the test of the past seven years, and *Opportunities in the Hydrologic Sciences* continues to function as a touchstone for the future of hydrologic science.

1

Wolman Lecture:
Hydrologic Science . . . in Landscapes . . .
on a Planet . . . in the Future

Thomas Dunne
School of Environmental Science and Management and
Department of Geological Sciences
University of California, Santa Barbara

BACKGROUND

In 1991 the National Research Council's Committee on Opportunities in the Hydrologic Sciences (COHS or the committee) proposed that there exists, or ought to exist, a distinct geoscience referred to as hydrologic science. Hydrologic science would be analogous to atmospheric science, geologic science, or ocean science but different from traditional hydrology, which the committee equated with engineering or applied hydrology. Both the aims and practice of the newly defined science were to be different from traditional hydrology. The goal of this paper is to examine whether this fledgling science has taken flight; whether it really has become distinct; and, if so, what it needs to sustain flight.

The committee adopted and elaborated on an Earth-science-based definition of hydrologic science, originally proposed by Meinzer (1942) and modified by the Ad Hoc Panel on Hydrology (1962):

> Hydrology is the science that treats the waters of the Earth, their occurrence, circulation, and distribution, their chemical and physical properties, and their reaction with their environment, including their relation to living things. The domain of hydrology embraces the full life history of water on the Earth.

This definition reflected developments that had been quietly occurring on several continents since the International Geophysical Year (1957-1958) and that presaged the International Hydrological Decade (1965-1974) and the continuing International Hydrological Program sponsored by United National Educational,

Scientific and Cultural Organization (UNESCO). These programs emphasized the study of hydrologic processes at all scales, the role of water in global-scale environmental processes, the need for regional-scale hydrologic analyses, the importance of hydrochemical processes, and the role of humans and other creatures in the hydrologic cycle. Because geophysicists and geographers were instrumental in developing these programs, and because of the breadth of those two fields as practiced in Europe, the research was seen as relevant to social concerns. It often assimilated hydrological research by engineers and foresters but was free from the pressure of immediate problem solving and design. However, these activities were not widely recognized by most hydrologists and were not strongly coordinated and taught as a coherent science.

The committee concluded that, although fundamentally an interdisciplinary activity, hydrologic science, being concerned with continental processes and their participation in the global water balance, should be viewed as a distinct geoscience interacting on a wide range of spatial and temporal scales with the oceanic, atmospheric, and solid-Earth sciences. Exploring these interactions is fundamental to understanding the behavior of water and the materials that it transports. Two points concerning the coherence of this scientific activity were left unresolved by the committee's suggestions: (1) a large number of scientists who study hydrology participate mainly through groups and funding agencies concerned with hydrometeorology and view themselves primarily as atmospheric scientists, and (2) there is still some question about whether hydrogeologists will integrate their activities within the vision of the distinct hydrologic science proposed by the committee or find it more attractive to act mainly within forums of geology and geophysics (Back, 1991).

APPLIED HYDROLOGY OR APPLICABLE HYDROLOGIC SCIENCE?

After reviewing the history of hydrologic applications, the committee argued for the existence and continued support of a hydrologic science distinct from traditional engineering hydrology. This is a point that needs to be carefully stated. The original distinction was not made to deny the value of problem solving or to judge the relative intellectual status of various activities. In fact, the COHS report was replete with promises of contributions to many societal concerns if hydrologic science were to be supported: ". . . the strengthened scientific base of hydrology will contribute directly to improved management of water and environment" (NRC, 1991). The report held out the promise of solutions to specific problems such as "the possible redistribution of water resources due to climate change, the ecological consequences of large-scale water transfers, widespread mining of fossil ground water, the effect of land use changes on the regional hydrologic cycle, the effect of nonpoint sources of pollution on the quality of surface and ground water at a regional scale, and the possibility of changing regimes of regional floods and droughts" (NRC, 1991). The committee argued, however, that historically this prospect had been too immediate: "elaboration of the field, education of its practitioners, and creation of its research culture have . . . been driven by . . . engineering hydrology" (NRC, 1991). Klemes (1988) has also been adamant that scientific research in hydrology be kept distinct from the technological activity of using hydrology to manage water resources.

The committee recommended instead the study of hydrologic processes and patterns at a variety of scales. Such pursuits would be free from the pressure to generate simulation models with a weak conceptual base, represented by boxes and arrows in flow diagrams, or with parameter values that have been obtained in a cavalier fashion for immediate application in design or regulation. The committee's hope was a long-term strategy, promising that "basic science" would pay off. Since 1991, however, much has changed in the national perspective about what research should be funded. Although basic environmental science seems to have fared quite well despite dire warnings, there is concern on various sides of hydrologic science that the growing call for applications will unduly constrain the kind of research that is supported.

The distinction between traditional applied hydrology and the applicable hydrologic science promoted by the committee represents two subtly different presentations of hydrologic understanding. Engineering courses and texts, which gave most of us our introduction to hydrology, tend to present the subject to users (decision makers, designers, students) as being complete or at least sufficient for action. This tends to focus attention on what is known or at least agreed upon, such as the hydraulics of simple channels or the physics of flow through homogeneous soils. It is natural in such an activity that one tends to search for the simplest definition of a problem to be solved or a task to be accomplished. A typical engineering hydrology task would be to predict flow depth, velocity, and overbank discharge for the design of a river-dredging project that will not cause undue harm to water quality or riparian wetlands (the latter being a relatively modern concern). This approach, focusing on one or a few aspects of the environment and adjusting to them or bringing them under control, has spectacularly advanced human health and welfare. Unfortunately, during the impressive history of this problem solving, society has sometimes changed its mind about what the problem to be solved is and has come to expect more sophisticated, higher-dimensional solutions to its needs and desires. For example, the simplification or destruction of aquatic and riparian ecosystems that accompanied many engineering activities formerly admired by society is now generally thought to be undesirable and in need of reversal (as long as the projects retain their capacity to support our material needs, of course!). Also, at least a few of the opinion-making elements of society have realized there are some limits to the stress that environmental systems can sustain and still remain in a preferred state. Meeting these expectations often requires a broader analysis of the systems and their influences than was the case earlier.

Engineering hydrology also tries to put into the hands of regulators, developers, and their designers the tools (equations, mathematical models, computer codes) that allow objective and quantitative decision making. Raising the confidence of the user, rather than emphasizing uncertainties and digging for deeper forms of knowledge about mechanisms, understandably takes precedence in the mind of the creator or compiler, especially in cases where there is an economic incentive to provide the standard method. Over the past three decades, there have been attempts to make some of these tools more rigorous and mathematical but not necessarily more physically realistic. Alternatively, other hydrologists have tried to make analytical tools more realistic representations of environmental processes by incorporating spatial and mechanistic aspects. After an optimistic beginning, these trends have recently been characterized as uncritical and even misleading. A number of senior engineers have, perhaps unfairly, laid responsibility for both the trend toward elaborate mathematical analysis devoid of mechanism and the optimistic attempts at spatially distributed simulation modeling solely at the door of engineers (Klemes, 1982, 1986; Nash et al., 1990). In fact, responsibility for unwise over-extensions could be laid at many feet. Nevertheless, it is said repeatedly by experienced practitioners (Beven, 1987; Beven et al., 1988; Loague, 1990; Grayson et al., 1992) that uncertainties in physically based simulation modeling are large and, when applied to design, planning, regulation, and other decision making, tend to mislead the consumers of such modeling results. Beven (1987) has even stated that spatially distributed, physically based hydrologic modeling is ripe for some kind of intellectual crisis and revolution that would redefine its scope and realistic possibilities. Similar comments on the misleading aspects of probability analyses, largely resulting from incorrect understanding on the part of users, have been made by Klemes (1982, 1989) and Baker (1993). There is need for a broader discussion of the goals of hydrologic modeling and of the relationship between model construction and empirical investigation than one currently finds in engineering hydrology alone.

The alternative presentation of hydrologic understanding, offered by COHS, is that of a science that confronts and even gains energy from its own uncertainties. In other words, it constantly focuses attention on what is not known and emphasizes the need for empirical exploration and explicit attempts to falsify or validate ideas. Such an activity seeks fundamental knowledge of natural processes, beyond the level required to solve a specific problem in construction, environmental management, or regulation. Freed from immediate problem solving and able to continually assess its own progress, the activity aims to construct and improve a coherent body of general theory. It also profits from lateral perspectives into ancillary sciences, looking for combinations of knowledge or analogous approaches. Thus, it is more likely to define questions about the operation of hydrologic systems in broad multidimensional terms. Activity of this kind has a good track record of discovering phenomena and insights that can be used to anticipate or solve problems in unexpected ways.

The theme of this paper, then, is that there is value in fostering a distinct hydrological science but that it will remain vital only if

• it discovers new phenomena, processes, or relationships governing the behavior of water and its constituents;
• it focuses on real hydrologic phenomena, such as floods, droughts, drainage basins, material storages and fluxes, and even large-scale engineering effects such as streamflow modification, soil conservation, or channel modifications; and
• it communicates itself well, both internally to develop cohesion and record progress and externally to develop support.

We should concern ourselves with phenomena that are of some interest to society, even if society must be continually informed about the significance of our research targets. The challenge at present lies in defining ways in which hydrologic scientists might gain support for their activities through functioning as engaged members of society. It is neither wise or even satisfying for hydrologists to develop a welfare mentality in which we expect that our research should be supported simply because we exist.

We also need to avoid the delusion of what Petroski (1997) has called the "linear" model of research and development in which it is assumed that basic research is the only precursor to intelligent problem solving. He recounts many examples of practical solutions to engineering needs preceding and even provoking fundamental understanding. The large-scale support for unfettered, curiosity-driven research promised by Bush (1945) seems in hindsight to have been largely a myth, popularized mainly by scientists who have enjoyed considerable freedom under an umbrella of funding provided subtly by some national concern for military, economic, health, or environmental security. Perhaps a more useful model for a scientist's career would be to see oneself as operating in a "web" of information arising both from nature and from society's interests. Keeping oneself broadly informed about nature, and about social concerns and needs related to the environment, can stimulate ideas and feedback and provide opportunities for gaining support, testing ideas, and contributing to human welfare. Thus, the desire for "freedom" from the pressures of immediate problem solving needs to be tempered with an acknowledgment of the value of stimuli that arise from the practical needs of society. This is true even when those needs are as diffuse as the needs of economists and policy makers to understand the limits of our ability to predict the consequences of large-scale changes in climate or land cover.

The following anecdote provides an encouraging example of how a pragmatic concern with water engineering has yielded fundamental hydrologic science of the caliber that any scientist would honor. In the 1950s the water supply engineer Law (1956) pointed out that Britain's post-World War I strategic need to establish a timber supply by planting conifers on the uplands of northern Britain would conflict with water supply to the rapidly modernizing industrial cities of England because increased canopy interception and evaporation would reduce the yield of runoff into reservoirs. The initial response from foresters was negative. But Law's (1957) field measurements strengthened his argument, which was taken seriously enough by government resource managers and hydrologists to gain their support for one of the most successful long-term paired-catchment experiments documenting land cover effects on water yield (Calder and Newson, 1979). The experiments were illuminated by studies of the physics of plant-water interactions (Rutter, 1967; Calder, 1990; Shuttleworth, 1989, 1991) that are still in operation. Such experiments have proven useful in analyzing land-atmosphere interactions relevant to both river runoff and global climate processes (Shuttleworth et al., 1984; Shuttleworth, 1988; Gash et al., 1996).

The distinction between what is fundamental and enduring and what is immediately useful in hydrology is further shaded by Black's (1995) illumination of the critical importance of "unused" resources. Black points out that, although an individual needs a relatively small amount of water to survive physiologically, much larger amounts are needed per capita to survive as a community and that all of Earth's water is needed to buffer the conditions that allow us to survive as a species. Understanding the state and functioning of water at any of these scales and the relationships between the various scales can be a useful contribution to society, if findings are translated into a form that is accessible to other members of that society. Engineering hydrology was not developed to analyze large-scale processes with multiple feedbacks between various loops in the water cycle. The current large effort led by the U.S. government to explore global sustainability surely ought to include a greater component of hydrologic science, even though it goes unrepresented on the Committee on Global Change Research of the National Research Council (1995).

Upon review this distinction between engineering hydrology, with its emphasis on immediate applicability, and a more measured but still pragmatic hydrologic science may seem artificial and unnecessary. However, it has a powerful influence on how hydrology is taught and practiced. Although it is not impossible to combine both aspects in a career, a graduate education program, or a government agency, there are important differences of emphasis. Most participants will become involved in one activity or the other, rarely crossing the boundary between approaches. There is even considerable suspicion across the cultural divide. Consequently, it is important to emphasize that a distinctive contribution to society can be made by harnessing the power of science in the analysis of hydrologic processes in ways that are not usually brought to bear in most engineering applications.

IMPETUS FOR HYDROLOGIC SCIENCE

The impetus for conducting a distinctive hydrologic science is the persistence of important gaps in society's knowledge about the roles of water in the operation of the Earth at all scales from global to soil profile. These roles are so multifaceted and complex that they cannot be elucidated by the narrowly focused approaches traditionally used in hydrology and taught in hydrology courses and texts, which have emphasized rather uncritical "can-do" approaches to the analysis and solution of water control and environmental problems. Although I have argued above that scientists should never forget their responsibilities, I remain supportive of the original COHS recommendation that there is reason to foster a unified discipline of hydrologic science, distinct from immediate design and regulatory needs. The distinctive requirements of such a science are that

- it seeks to identify and study fundamental processes, sometimes going beyond the needs of the immediate task or the community interest of the moment;
- it explores connections to collateral influences on the behavior of water, such as the role of biota, the solid Earth, or the oceans;
- it aims to construct a coherent body of transferable theory; and
- it is obligatorily self-critical, taking seriously the importance of falsification through critical measurements.

There is great value in fostering such activity, both because it is the interesting and distinctive option in hydrology at this time (Klemes, 1986; Beven, 1987; Dooge, 1988) and because such an activity can contribute to understanding problems of keen societal interest and utility, now and in the future. This opportunity can be illustrated with two examples, both drawn mainly from surface hydrology, that is close to my own experience. Although equally interesting contributions are being made in hydrometeorology, hydrogeology, and biogeochemistry, it is easier to make this point with familiar examples. I argue later that, although the promise of making useful contributions to knowledge exists, the continued vitality of the fledgling hydrologic science is not assured, and it needs some attention from those who place value on it.

The two examples derive from assertions that may seem whimsical at first glance. A student will find no more than a passing reference to them in even modern hydrology texts. Yet they are two of the most important realizations in hydrology in the past two decades. The assertions are as follows

- We live on a planet, and that fact has hydrologic significance ranging from the mechanisms by which the energy and water balances of the entire Earth are stabilized in a range that is hospitable for humans and other biota to investment decisions that must be made by local water authorities about their future water supplies. Furthermore, we now realize that human activities can influence

the hydrologic cycle on this scale by altering the chemistry and radiative properties of the atmosphere and possibly by transforming the Earth's land cover on a
sufficiently large scale. There is considerable interest in anticipating, or at least
defining, the uncertainty about the impact of global change on the hydrologic
cycle and water resources.

  • We live in landscapes, the topography of which is a dominant influence
on spatial and temporal patterns of water storage and surface and subsurface
transport of water and its constituents. We also recognize that humans can
influence the hydrologic cycle on the scale of even the continental-scale river
basins. This is a consequence of the spatial extent of land cover changes and the
degree to which humans have intensified or perturbed biogeochemical cycles
through pollution and extensive land transformation.

Both of these insights present enormous challenges to hydrologic science
and simultaneously provide it with opportunities for contributing to human welfare and reducing ecosystem disruption. They do, however, require us to change,
or at least expand, our hydrologic science.

SIGNIFICANCE OF PLANETARY–SCALE HYDROLOGY

Asrar and Dozier (1994, p. 6) describe the Earth system as "two subsystems—
physical climate and biogeochemical cycles—linked by the global hydrologic
cycle." Few traditional hydrologists would have thought of elevating the significance of the hydrologic cycle to this level. However, the implied challenge is a
measure of the degree to which hydrology has recently been called on, even
entrained, by atmospheric scientists to answer questions about the linkage between land surface processes and the atmosphere.

Hydrometeorologists seek collaborations to understand the land-atmosphere
interactions that influence tropospheric circulation on all scales, particularly the
global and regional redistribution of water and heat. Shuttleworth (1988) and
Sellers et al. (1997) summarize recent developments and outstanding questions in
this field. In order for them to improve their models of land-atmosphere interaction, hydrometeorologists need to refine their knowledge of spatial and temporal
patterns of moisture storage and availability for evaporation. Therefore, they
need to know about the distributions of soils, plants, and topography and their
roles in holding moisture or releasing it to deep ground water or streamflow.
Typical wavelengths and amplitudes of topography in various physiographic
regions, as well as regional patterns of soil and plant distributions, affect the rate
at which water drains from a landscape and therefore the amount and pattern of
its storage and availability for evaporation. Although there have been many
illustrations of spatially distributed hydrologic modeling, most of them have been
concerned with outlining model structures and demonstrating that they can be

calibrated to match streamflow responses. There has been little use of thoroughly substantiated models to explore the effect of first-order geographical patterns in a way that has some theoretical relevance and widespread applicability, even if only in an approximate way at this time. Eagleson's (1978) papers indicate the combination of rigor and intellectual reach that is needed as a starting point in this field. Blöschl and Sivapalan (1995) and other contributors to the same volume review the hydrologic significance of spatial patterns of soil and topography, and Wood (1995) illustrates how a combination of modeling and remote sensing could yield information about the effect of measurement resolution on computed results.

An important challenge for improving our knowledge of hydrology at these continental scales is to better represent processes, material properties, and boundary conditions that are characterized by such small-scale spatial and temporal variations that they cannot be resolved with foreseeable measurement and computational resources. Such processes and properties have to be represented in models through the strategy of parameterization, which expresses the averaged behavior of these unresolvable effects and their multifarious nonlinear interactions on process rates (e.g., the average rate of evaporation, average rate of erosion, or average speed of water evacuation from a landscape). These effects are often presented as a nuisance that hinders prediction because of "parameter uncertainty." Viewed from another perspective, however, they present a focus for investigating characteristic patterns of material properties, of topography, and of the processes themselves, leading to hydrologic discoveries. There is much new science to be done in investigating these patterns and in discovering new aspects of the behavior of water and the materials it carries, rather than focusing most of the discipline's attention on improving computational methods and on calibration of simulation models. Learning how to represent these patterns of materials and processes offers the prospect of improving their model formulation in ways that will truly enhance our understanding of the hydrologic cycle.

Encouraging possibilities exist for new forms of hydrologic measurement, taking advantage of the revolutions in electronics and particularly remote sensing. However, these advances are not panaceas. Satellite-based remote sensing yields a pixel-averaged view of the Earth's surface from the top of the atmosphere, with varying degrees of spatial resolution depending on the sensor. In general, remote sensing is better at providing spatial coverage than the temporal coverage valued by small-scale hydrology. We must strive to combine remote sensing products with more thorough and creative fieldwork in order to investigate what is being represented. It is difficult to imagine the range of opportunities offered by satellite-based remote sensing because of the rapid evolution of this technology (Asrar and Dozier, 1994). Much work remains to be done both in interpreting and using remote sensing products for hydrology and in articulating the specific data types and resolutions needed to solve critical hydrologic problems. That work is gradually developing into field campaigns for coordinated measurement and, more slowly, into the development of a theory-based consensus about what spatial and temporal patterns of properties and processes need to be investigated.

The other side of this emergent interest in planetary and continental-scale hydrology is the need to interpret the hydrologic consequences of any anticipated combinations of climate change and other large-scale environmental change, such as population growth and deforestation. Society needs an improved capacity for predicting the status of water resources (sensu lato) on seasonal-to-interannual time scales that immediately influence human activities. In addition, the results of global and regional models of change (atmospheric general circulation models [GCMs], demographic projections, land use scenarios) need to be translated into predictions of ground-level conditions and processes such as soil-moisture regimes, ground water recharge, runoff volumes, floods, droughts, lake levels, and soil erosion patterns. This effort has begun (Wigley and Jones, 1985; Wolock and Hornberger, 1991; Marengo et al., 1994) but is stymied mainly by uncertainties in GCM-predicted magnitudes or even signs of changes in annual and seasonal precipitation and radiation loads and the need to interpret precipitation event and plant characteristics to be expected under altered climatic regimes. Uncertainties about terrestrial factors can be partially overcome by cautious space-for-time substitutions that introduce reasonable scenarios of expected changes, even if they cannot yet be predicted from first principles. Such scenarios are more likely to be accurate if generated by experienced field scientists with a theoretical turn of mind.

The problem of anticipating global warming effects may be a little easier for river basins in the snow zone. Lettenmaier and Gan (1990) and Nash and Gleick (1991) have used a simple model of snowmelt and the resulting basin runoff to analyze the sensitivity of snowpack and soil moisture storage and of seasonal and flood flows to various climate scenarios generated by two GCMs for hypothetical enhanced atmospheric carbon dioxide conditions. The results suggest that significant changes should be expected in the seasonal timing of runoff—that is, more runoff in autumn and winter because of lower ratios of snow to total precipitation and lower spring-summer runoff because of lower snowpack storage. Such changes would have expensive consequences for the kind and amount of artificial streamflow regulation that might one day be needed. However, much better spatial and temporal resolution is required in both the atmospheric and land surface models before much confidence can be placed in anything other than the sign of the results. The easily available models, originally developed for calculating the water balance of individual soil profiles or for quasi-black box forecasting of floods, are simply too crude to rely on the quantitative results. However, the results are convincing enough to give some urgency to the task of improving modeling capability for large river basins. They may also compel society to seriously consider the possibility of major disruptions of its water storage, supply system, and flood risk in the snow zone. In addition to model developments, uncertainty must be reduced concerning representative field conditions in basins typical of the physiographic region. These include the amount of ground water storage, the interannual carryover of water, and the influence of elevation effects on snowpack accumulation and the energy balance. A combination of field surveys and remote sensing of the current range of conditions will be needed to allow model-based extrapolations into unsampled environments. This can best be accomplished with coordinated empirical and modeling studies. Throughout such an effort, however, hydrologists must be responsible for explaining to people the significance of even preliminary results and why such an investment in measurement is required to gradually reduce uncertainty.

By comparison with the climatological and meteorological uncertainties, differences of emphasis among hydrologists about modern process-based models and the difficulties of parameterizing them seem small. Beven et al. (1988), Wood et al. (1988), and Robinson et al. (1995) among others have drawn together the current level of field experience into a convincing theory of how hillslope and channel network characteristics govern basin hydrologic response, including the distribution of soil moisture at a range of geographical scales. This statement is not meant to ignore the massive problems introduced by both systemic and parameter uncertainty in process-based modeling of surface hydrology (Beven and Binley, 1992; Grayson et al., 1992;Beven, 1993; Jakeman and Hornberger, 1993). However, some of those problems arise from a lack of attention to uncertainties that can be reduced with field measurement or from asking intractably difficult questions in the first place. The optimal recursive blend of computation and field measurement is rarely explored in answering broad, socially relevant questions in hydrology. This situation needs to be changed, possibly through strategies suggested at the end of this paper.

Appreciation that planetary-scale atmospheric changes might affect continental climate and hydrology has also resulted in the resurgence of empirical hydroclimatology. Redmond and Koch (1991) have investigated the statistical association between indices of large-scale atmospheric circulation, seasonal precipitation, air temperature, and regional patterns of streamflow during the past half-century. They showed that winter precipitation (October-March) and annual streamflow are generally low in the Pacific Northwest during El Niño-Southern Oscillation events. These conditions involve a weakening of atmospheric pressure gradients across the Pacific, a weakening of easterly winds along the equator, and higher-than-average sea surface temperatures off South America. Equatorial sea-surface temperatures strongly affect the transport of heat, pressure distribution, and wind patterns in midlatitudes. Values of the pressure anomaly indices known as the Southern Oscillation Index and the Pacific/North America Index during the preceding June-November period correlate with Pacific Northwest rainfall during October-March. Winter air temperature, on the other hand, is negatively correlated with this atmospheric index. Drier-than-average winters thus tend to be warmer than average. This association enhances the variability of snowmelt contributions to streams.

Cool wet winters provide deep snowpacks, whereas the thinner packs accumulating in warm dry winters tend to melt faster during winter and be unavailable for supplying summer runoff. Annual streamflow was also correlated with the Southern Oscillation Index for the June-November period preceding each water year with generally congruent results: during El Niño events, below-average flows can be anticipated throughout the northwestern states. Opposite results occur in the rainfall and streamflow of the desert southwestern region. Cayan and Peterson (1990) found similar positive correlations between the winter index and streamflow in the succeeding half-year in these two regions. Webb and Betancourt (1990) uncovered the association between intense El Niño events and floods in succeeding winters in the desert southwest. Thus, precipitation, air temperature, and streamflow are significantly related to atmospheric circulation patterns occurring thousands of kilometers away and several months before the high-flow season. They vary in a manner that produces alternating runs of wet and drought years and perhaps lower-frequency variability that has not yet been recognized in the relatively short records of atmospheric circulation indices. Whalton et al. (1990) have used even more extensive, though sparser, records to demonstrate large-scale patterns of climate state and their influence on hydrology around the Indian Ocean basin.

These associations raise the possibility of at least short-term predictability of regional precipitation and streamflow as models of atmospheric and ocean circulation are improved. Although the statistical associations are not very precise at present, there could be enormous economic implications of even weak correlations for improved reservoir operation for hydroelectric generation, flood control, and other purposes; the prediction of region-wide streamflow in salmon-bearing streams; and even the magnitude of fish runs several years later (because of the effects of a year of high smolt production on the numbers of fish returning to spawn several years later).

Public policy at all levels of social organization now requires hydrologists to interact with atmospheric scientists, biogeochemists, and others to gradually reduce uncertainty about the terrestrial and atmospheric processes at planetary and regional scales that may be affected by natural and anthropogenic environmental change. This is such a challenging field in which to create true breakthroughs in understanding that it will require a long-term commitment to gradually uncovering the tractable questions, developing better ways of making measurements, and staying abreast of developments in atmospheric circulation modeling. Thus, it will require the kind of broad scientific training envisioned by proponents of a modernized hydrologic science.

SIGNIFICANCE OF LANDSCAPE HYDROLOGY

One of the earliest formal analyses of the role of topography on hydrology at whole-landscape scales was the regional flow net theory proposed by Toth (1963, 1966). Topographic features of various scales from hillslopes to ridges (Meyboom,

1962) and up to mountain-lowland juxtapositions (Duffy, 1988) drive ground water circulations of various sizes, residence times, response times, and chemical and physical properties. The extent of these circulations, their interdependence, and the transition zones between them could be anticipated, documented, and interpreted. The conceptual and mathematical model, which Toth solved analytically for the simplest case of sinusoidal topography over a homogeneous isotropic aquifer, allowed major breakthroughs in understanding the behavior of ground water systems. It also provided a guide for the application of numerical models that were soon to transform subsurface hydrology.

At the surface, landscapes consist of complex hillslopes converging on channel networks that in map view have a treelike hierarchical structure. Topography creates preferred pathways for overland and shallow subsurface water (Kirkby and Chorley, 1967; Dunne et al., 1975; Beven and Kirkby, 1979) and for sediment (Dietrich and Dunne, 1978; Benda et al., in press) directly through the gravitational effect and indirectly by creating patterns of soil properties such as depth (Moore et al., 1993; Dietrich et al., 1995), macropore concentration (Ziemer and Albright, 1987), and rooting strength (Crozier et al., 1990). The transition from convergent portions of hillslopes to channels is a crucial process threshold (Dietrich and Dunne, 1993) affecting the evacuation rates of water and sediment. The density of channels per unit area is an important landscape characteristic that depends on climatic and geotechnical properties of the terrain (Horton, 1945; Abrahams and Ponczynski, 1976; Dunne, 1980).

As channels converge and enlarge downstream, they widen their valleys and create floodplains in which large volumes of water and sediment can be stored, although the time constants of storage for these two materials are radically different, being days to months for water in a large valley floor and hundreds to thousands of years for sediment (Dietrich and Dunne, 1978; Trimble, 1983; Meade et al., 1990; Mertes et al., 1996). Potter (1978) has interpreted how the continental-scale alluvial river valleys are localized in the context of global tectonics. In addition, empirical studies of the sediment budgets of large rivers and their extensive valley floors are being investigated for natural (Kesel et al., 1992; Dunne et al., 1998) and polluted-sediment cases (Lewin et al., 1977; Marron, 1992; Graf, 1994). Much work in geomorphology is currently focused on understanding the relationship between the storage of water and sediment in valley floors and the flooding regimes and morphology of the alluvial environment (Gomez et al., 1995; Mertes, 1997; Nicholas and Walling, 1997).

There is, in other words, a coherent qualitative theory of the topographic control of water and sediment movement at the scale of whole landscapes, even up to those of regional scale. New methods of tracing and dating deposits and morphological change have even allowed quantitative interpretations to be made and used for decision making. These landscape-scale theories of basin runoff, material transport, valley floor geomorphology, and inundation mechanisms have direct relevance to attempts to build landscape-scale theories in aquatic ecology (Vannote et al., 1980; Junk et al., 1989; Stanford and Ward, 1993; Power et al., 1995).

Quantitative process-based models for whole drainage basins or regional landscapes that might be used as a basis for reducing uncertainty in decision making are much more fragmentary. Ahnert (1976), Kirkby (1986), Willgoose et al. (1991), and Smith et al. (1998a, b) have modeled the formation of channel networks and drainage basins as the result of hydrologically driven sediment transport. The models reveal how landscape properties reflect the climatic regime, material properties, and geometrical constraints such as tectonic deformation and baselevel change. Beven (1986) elaborates upon the basic TOPMODEL quasi-steady-state approximation for both overland and subsurface runoff from whole catchments. This approach is now widely applied and incorporated into interpretations of large-basin flood runoff (Miller and Kim, 1996) and the surface hydrologic implications of global climate change (Wolock and Hornberger, 1991). As with all other spatially distributed environmental models, the approach is difficult to validate because of constraints on our ability to map transient hydrologic characteristics, but the results agree in a qualitative way with much field experience.

In the analysis of hydrologic response in all but the smallest river basins, a problem results from the large number and diversity of hillslopes and responsive swales (convergent portions of the topography) in a basin. These source areas for runoff possess frequency distributions of topographic characteristics (e.g., gradient, length, concavity), hydraulic properties of soils, and plant characteristics. Their hydrologic response is stimulated by rainstorms that are themselves discrete in both space and time but have definable probability distributions of these hydrometeorological characteristics. Thus, stochastic approaches to runoff generation were proposed to capture the potential variation of responses within a basin and their aggregate behavior, which could then be used to derive a response function that could be routed down a channel network. Basin hydrologic response to storms could then be derived from hillslope runoff and channel network structure (Rodriguez-Iturbe and Valdes, 1979; Gupta et al., 1980). This approach, named the Geomorphologic Instantaneous Unit Hydrograph, provided one of the most promising, simplifying insights in attempting to build a theory of basin response. Robinson et al. (1995) have examined the respective influences of hillslope processes, channel routing, and network geomorphology on the hydrologic response of natural catchments. Benda et al. (1998a) and Benda and Dunne (1998b) have developed a similar stochastic approach to the generation and routing of sediment from drainage basins, examining the influence of basin size, erosion processes, rainstorm and fire climate, and other controls on regimes of sediment transport and storage.

As in the case of continental- and planetary-scale hydrology, all of these attempts at theory building are limited by difficulties in representing subgrid-scale processes, measuring the spatial variability of landscape properties, and measuring the temporal characteristics of climatic and hydrologic responses over entire drainage basins. In some environments and at some scales there also exist continuing systemic uncertainties about the physics and chemistry of the mechanisms themselves. Fortunately, the systemic uncertainties are probably reducible to a significant degree through continued emphasis on programs of field observation. In addition, improvements in technology can probably assist in characterizing the variability because the sources of variability themselves are quite well understood (Seyfried and Wilcox, 1995). However, the complexity of the mechanisms and the intensity of property variation require an approach that in other fields is called "coarse graining" (e.g., Gell-Mann, 1994, p. 29). The goal is to formulate scientific problems with only the necessary degree of complexity, using simplified expressions to represent processes and their interactions. When formulating a particular problem, the scale must be chosen judiciously to maximize its utility. This challenges the belief that only progressively finer-scale studies are really "scientific" and "rigorous" and that "scaling up" from some presumably fundamental understanding is the only way to solve problems in hydrology. As we try to "scale up" our understanding of many hydrologic phenomena, either in formulating theory or in interpreting measurements such as satellite products, it appears that different processes exert their complicating influence at each scale. At some point we encounter the current limits of our ability to formulate, measure, or compute, and we resort to the strategy of parameterization of subgrid effects, as discussed above. Once again, at the landscape or basin scale, this need for intelligent and thorough parameterization provides a focus for integration of theoretical and field investigations.

Problems of measurement also account for the difficulty of verifying any of these theories quantitatively. For example, the models used for illustrating concepts of catchment runoff response summarize extensive field experience; they calculate the consequences of accepting various postulates about mechanisms and parameter values that they include. However, as Beven (1989, 1993, 1996) and other experienced modelers have often repeated, the quantitative results that the models provide are highly uncertain because of parameter uncertainty and in some cases the inaccuracy of current modeling concepts. Yet they are often mistaken for reliably precise prediction methods by those looking for a convenient, apparently objective way of making decisions about environmental management. Model capabilities are frequently overstated, or at least not critiqued, by their developers and other proponents. Much self-critical work needs to be done to define realistically the levels of uncertainty in model results. This is necessary if, in the face of such uncertainty, we are ever going to be able to provide scientific advice on how to make environmental management decisions about, for example, whether extensive timber management has long-term consequences for flood levels (Jones and Grant, 1996).

Given the importance of regional-scale river basins and their valley floors as loci of settlement, agriculture, commerce, and ecological values, it is surprising that so little theory and observation have derived from the major floods that bring both catastrophe (Chagnon, 1985) and productivity (Junk et al., 1989) to the river corridors of large basins. At drainage areas greater than about 10^4 km^2, physical theory in flood hydrology seems to be abandoned in favor of statistical analysis of floods as random variates. Klemes (1982, 1989) and Baker (1993) have forcefully argued that this was never a good idea, given such well-known difficulties as hydrologic persistence, the variety of flood-generating meteorological events, potential anthropogenic influences, and the manifest nonstationary behavior of the few century-long flood records. The strategy is even more likely to be misleading when society is facing uncertainty about the effects of environmental change on floods and droughts. The technique of estimating the probabilities of future floods and droughts based upon an analysis of the historical record of discharge extremes may have been reasonable in the early days of engineering hydrology, when engineers needed realistic, if approximate, estimates of the risk to projects. It is also understandable that the convenience of the method, and the need for a lingua franca with which to express risk to regulators and investors with an actuarial point of view, would lead to a high level of mathematical development of this field of statistical hazard estimation (Maidment, 1993). However, says Klemes (1989, p. 43), "the main weakness of the analysis is that it takes no account of the actual climatic, hydrological, and geophysical mechanisms that produced the observed extremes." A focus is needed on the processes that actually generate floods, including extreme runoff generation and therefore large rainstorms, snowmelt rate and timing, the slow drainage of previously wet soils, land cover changes, land drainage, and even dam failures or operational inefficiencies.

In recent decades there has been little quantitative study of floods as processes or of how the process of flood generation is influenced by large-scale environmental controls, especially in large basins. Although attention is now being paid to the mesoscale atmospheric circulation processes that govern rainfall character, there is no comparable ground-level investigation of the flood-generating role of season-long soil moisture patterns, regional-scale patterns of soil waterholding properties and topography, the hydrographic structure and disparate geography of major tributary basins, valley floor geometry, regional land use change, land drainage, or diking over 1,000-km scales, as occurs along the Mississippi River and is proposed in the Bangladesh Flood Action Plan. Mertes (1997) and others are only just beginning to reevaluate the complexity of valley floor inundation. From the debates following the 1993 Mississippi floods, there appears to be little quantitative consensus about the role of valley floor storage in modulating the peak discharge of large protracted floods except near levee breaks. Aside from direct flooding hazards to the habitability of valley floors, we have little quantitative understanding of the hydrogeology of valley floor drainage and its role in wetland distribution and functioning or of floodplain sedimentation (Gomez et al., 1995; Nicholas and Walling, 1997) and the disposition and recruitment of contaminated sediment and water in populated floodplains (Marron, 1992; Meade, 1995). All of these are process-related issues that could be the focus of significant research aimed at improving the habitability of alluvial lowlands, now that large parts of them are irreversibly developed.

The 1993 floods in the Missouri and Mississippi basins, which caused approximately $12 billion in property damage (Myers and White, 1993) as well as other monetary losses, massive water pollution, and damage to dikes, generated an extensive but short-lived debate about how the nation should respond to the prospect of future similar inundation. However, the debate was not guided by hydrologic science. In fact, in the four years since the nation's largest hydrologic event, one can find virtually no analysis of the flood in the several front-rank scientific journals in hydrology. Nor have these journals published any fundamental reanalysis of the nation's options and opportunities for creative responses to future events, either immediately after the flood or in succeeding years when there was time for more leisurely analysis of the runoff generation and storage processes governing the magnitude of the flood. By comparison, the equally momentous 1980 eruption of Mt. St. Helens generated significant scientific field investigation and data collection that was assimilated by the geological and geophysical community and later applied to guiding the social response to eruptions at Redoubt and Pinatubo volcanoes. After the Mississippi flooding, federal agencies compiled data reports, but the hydrologic scientists seem to have ignored the largest hydrologic event in this nation during their careers. There was, however, the usual debate about the computed recurrence interval!

CAN HYDROLOGIC SCIENCE TACKLE COMPLEX LARGE-SCALE PROBLEMS?

The types of problems outlined above are complex, multidisciplinary, and challenging, and even the definitions of some of them need refinement. Those outlined are not even the most complex problems for which society could use solutions. The question naturally arises of whether they are too complex to be tractable at any fundamental level that would be called scientific by some of the critics of hydrology (Dooge, 1986; Klemes, 1986) or whether they lie in the area of "transcience" (Weinberg, 1972).

It could be argued that when some form of objective calculation is necessary we should rely on calibrating low-resolution models for large-scale hydrologic phenomena and concentrate studies of "fundamental" processes and complicated calculations at small scale. But the relevant processes are not limited to small-scale phenomena. The gathering of a flood in a continental scale river valley is a process. So is the impact of regional drought on the outcropping and discharge of ground water in a stream network or the down-valley advection and diffusion of a sediment wave released from hillslopes by agricultural colonization. Second, though it may be difficult, improving the analysis of large complex problems is an extremely interesting option for hydrologic science at this time. Examples include understanding the hydrologic significance of persistent planetary-scale atmospheric conditions or the consequences of perturbing the surface properties of regional landscapes.

Scientific analysis of complex large-scale processes can generate fundamentally new ways of looking at hydrologic change, require new modes of observation, and stimulate new questions for analysis at smaller scales. It requires some judicious application of the coarse-grained approach, referred to above (Gell-Mann, 1994), so that integrative theories may be developed similar to those of Toth (1963), Beven and Kirkby (1979), or Rodríguez-Iturbe and Valdes (1979). Such an exercise may force us to conclude that certain questions about small-scale hydrologic processes have been adequately researched for the present and are not likely to yield progress until someone produces a new integrating concept. In surface hydrology there is, for example, a particular need for replacing storm runoff models based on the Richards equation with more realistic descriptions of flow in macropore-filled regolith, the large-scale geometry of which may have characteristic patterns within a drainage basin. Another example is the need to modify or even replace current models of sediment transport that were originally developed and calibrated in flumes to compute instantaneous transport under an infinite sediment supply. A more realistic view is that of episodic sediment transfer down river networks during floods of finite duration, during which the supply of sediment may be limited by external processes or the state of the channel bed.

Developing realistic integrative theories about large-scale complex processes, even if they are coarse grained, would allow hydrologic scientists to focus on research targets that are of broad significance and would attract sustained interest from scientists in other fields and society at large. Atmospheric scientists have pointed the way by studying how the energy and water balances of extensive continental surfaces affect atmospheric dynamics at regional and larger scales. That information is needed for reducing uncertainty in global circulation models, but once the activity grew it became apparent that it could also lead to improvements in local meteorology and hydrology. Such research activity arises within a field—in this case, atmospheric science—which has a strong theoretical tradition and therefore a basis for internal communication, agenda building, and the design and advocacy of short- and long-term data collection programs. Hydrologists would be well advised to emulate these skills from a close sibling group of colleagues who also work outdoors. A key to being able to tackle large complex problems appears to be the ability to forge a consensus approach between theoreticians and empirical investigators.

HAS HYDROLOGIC SCIENCE TAKEN FLIGHT?

Some recent commentators on the status of hydrologic research have referred to hydrologic science mainly in terms of its potential but have concluded that for the present the niche of geoscientific hydrology is essentially empty. Although such conclusions are meant to be taken as stimulation rather than condemnation, it is probably time, after a decade of such conclusions, to acknowledge that a review of the literature would leave one quite impressed. During the past decade, the output of all hydrologic journals has increased dramatically, and in the main research journals, such as *Water Resources Research, Journal of Hydrology, Journal of Geophysical Research, Reviews of Geophysics,* and *Hydrologic Processes,* many papers have appeared that would be classified as scientific hydrology by even the harshest critics of past practices. That conclusion can be extended to those journals to which hydrologists have contributed more recently, such as the *Journal of Climate* or *Earth Surface Processes and Landforms.* The problem is that the essence of hydrologic science as proposed by the COHS does not emerge clearly amid all of the other, albeit high-quality, papers on topics related to algorithm development, site characterization, and management.

Another measure of activity could be the types of hydrologic and related research funded by the Hydrologic Sciences, Water Energy, Atmosphere, Vegetation, and Water and Watersheds programs of the National Science Foundation between 1993 and 1997. Many fine ideas have been pursued, but they demonstrate more diversity than coherence. The Hydrologic Sciences program, in particular, has represented the breadth of opportunity expressed in the original COHS report that recommended its establishment. This is not to suggest that diversity is a bad thing for hydrology, merely that it would be difficult for an outsider to recognize the coherence of hydrologic science from the research projects funded so far by these programs. One could argue that, since it is wise for a fledgling science to invest in variety, the current low funding levels preclude the development of critical mass in particular foci. While this may be a valid point to some, it does not hide the apparent lack of a coherent agenda in terrestrial hydrologic science. By contrast, in the hydrometeorology component of hydrologic science, funded outside these programs, both a more coherent agenda and a higher level of funding are apparent. Presumably, more significant funding was not granted as a whim but in response to a well-defined, coherent, and skillfully articulated and publicized research agenda. The proposals caught the imagination of policy makers and budget managers, who could clearly see the social benefits or at least public interest in the scientific ideas being propounded. By contrast, the hydrologic science of the ground surface and subsurface, though visibly in flight, is flapping around in sight of many exciting possibilities, needing to catch a good breeze in order to sustain flight.

WHAT IS NEEDED TO SUSTAIN HYDROLOGIC SCIENCE?

Much excellent scientific hydrology is under way, and there are many attractive and socially useful scientific questions to work on, particularly at landscape and continental scales in addition to the hillslope, site, and water catchment scales that are already well represented. However, these questions about larger-scale processes will remain difficult to research in a scientific manner unless we can devise theories and measurements appropriate for those scales. To do so we need to solve some internal problems about the ways in which we interact in research and teaching and in how we communicate among ourselves and with society. I will close by making some suggestions for ways of overcoming some of our difficulties in this regard.

Convergence of Approaches

Dooge (1988, p. 79) has recommended that hydrologists should resist the fragmentation of approach to their subject. This point needs continual emphasis simply because it is so difficult to achieve and yet is an obvious impediment to the development of hydrologic science. Most of us enter hydrology from other fields, having no tradition of communicating except to sponsors and graduate school colleagues who identify with our particular interests in hydrologic contributions to flood management, water supply, soil science, water quality, or landscape evolution. We have learned empirical approaches in the field or laboratory, or mathematical approaches if that was more attractive, and we tend to search for problems, places, and sponsors that are amenable to the practice of our skills. Most of us are too busy or disinterested to broaden our skills in any radical way after graduate school, and we are not usually encouraged to do so. Thus, we become identified as a particular kind of hydrologic scientist: geomorphologist, soil physicist, ground water hydrologist, surface water hydrologist, etc. We attend different sessions at scientific meetings, and spend little of our time contemplating the connections between weather, surface water, soil hydrology, and ground water that—for example, at the scale of a large river basin—might hold keys to the generation of floods. We become identified as "modelers" or "field hydrologists," "theoretical hydrologists" or "just descriptive." Few of us, with some notable exceptions, struggle to avoid the ossifying identification, and it is not much of a step from there to begin denigrating other approaches in the competition for support of all kinds. Yet experienced scientists repeatedly emphasize that scientific breakthroughs commonly arise when scientists break out of their disciplinary isolation and collaborate in the unexplored territory between specialties.

Specialization is obviously required for skill building, so I am not proposing anarchy in the organization and conduct of the science. However, imagine that early in graduate school and through constant mentoring we were reminded that (1) we are not owed a living but should seek to contribute to human welfare if we want to be granted the good fortune of a scientific career, (2) we can harness the power of science to provide society with an intellectual compass for adjusting sustainably and productively to its environment, and (3) the scientific questions on which we might therefore work are not sensitive to disciplinary differences or status concerns among hydrologists. Perhaps then we could view the differences in our skills as resources rather than impediments to the development of more unified approaches to the study of water. Such a graduate school orientation would encourage even extremely specialized scientists to appreciate different approaches in related fields of hydrology, to take seriously the need for communication with other hydrologists, and to express the connections between their own results and other insights obtained at different scales and by different means.

Hydrologic scientists will continue to come from a broad range of backgrounds. Even if new graduate programs are developed in hydrologic science—which is an excellent idea that some universities can afford—most people entering the field will be trained and employed outside such programs for the foreseeable future. Back (1991) has even recommended that hydrogeologists resist being trained in a unified hydrologic science and instead stay grounded in departments of geological science. Whatever the mix of backgrounds, respect and interest must be built across them in order to encourage hydrologists to spend the time necessary to learn and unite the various forms of understanding that we develop. This requires communication that is not even widely valued at present. There will be hydrologists who are skilled at abstraction and the formalization of theoretical models. Such people are needed to see through the complexity and disparate impressions of our experiences and to develop useful generalizations. Other scientists are intrigued by particular observations, particular cases, or field observations of many disparate cases. They have seen a great variety of environments and measured hydrologic events or distributions, and they have good integrative skills, which gradually yield generalizations. They tend to be the ones who discover processes and relationships in real environments, but they do not spend much time on generalized theory, and some of them do not even value it.

We are fortunate to have this range of skills among people interested in hydrology, but it is unfortunate and debilitating that these individuals or groups rarely communicate and sometimes even demean each other's contributions. There are exceptions (Stedinger and Baker, 1987), but they are rare. University departments often choose to hire only one kind of hydrologist. Thus, their graduate students see only a sliver of the full range of approaches available in hydrology, and they build no skill in communicating with other kinds of hydrologists. This limitation will eventually reduce their satisfaction and even their job prospects unless they can remedy the situation during their careers. We have to overcome this problem in our education programs, in the conduct of our societies, and in the work of agencies that utilize hydrologic science.

We must find vehicles for combining various approaches. One possible mechanism is to focus on the construction of unified theories in hydrology. This will not be accomplished by a small group working alone indoors. It requires the interaction of people with a variety of experience who are prepared to spend time combining their information and talents. If the hydrologic societies cannot effect this through the repeated organization of integrative symposia or special sessions, perhaps it is time to take the example of theoretical physics, which for decades has utilized the summer institute as a means of bringing together groups to work for periods of up to three months on particular problems. These institutes require a culture of sharing information that has not yet arisen in all fields of science, including our own. Hydrologic science may not be ready for such a large-scale commitment, but it may be time to experiment with a limited form of the concept.

The ability to construct a coherent theory-guided agenda seems to be a key to the convergence of approaches advocated by Dooge (1988). Diversity of approach has great advantages, but it is important that we at least understand where our colleagues who utilize different approaches are located on the intellectual map of hydrology. It is a useful exercise for all of us, and particularly for the training of students, if we can articulate and at least partially defend the approach of groups who may be working in a radically different manner or direction from our own. Such activity could be a form of consensus building to guide hydrologic science. It will not suit the personality of every hydrologic scientist, some of whom we can count on to ensure a diversity of viewpoints. But if we are ever to conduct truly scientific investigations of large-scale complex hydrological processes of the kind described above, it will be necessary to marshal our investigations more thoroughly than is the tradition in hydrology.

Development or recapitulation of theory would be an important tool in agenda building for hydrologic science. The theoretical physics model referred to above suggests that it would increase and intensify collaborations and lead to proposals possessing a wide degree of reviewer support. That is the surest way to build the proposal pressure that would justify better funding levels. Strengthening of a coherent scientific agenda, rather than diffuse calls for more "scientific approaches," would attract students, sponsors, and clients. Such an agenda would include important research questions; plans for measurement programs, including critical tests of ideas; and proposals for linking findings about aspects of hydrology at different scales.

Communication

If scientists with different backgrounds are prepared to invest time in more unified approaches to hydrologic analysis, a means of communication will be needed to integrate them. The elements of communication needed for a robust hydrologic science must be provided in graduate programs of the science, whether they are coordinated in single departments or in federations of departments. However, the science cannot await only the training of a new generation of young people. We need to promulgate immediately certain kinds of communication and retooling among ourselves, using our professional societies, oversight bodies, and personal conduct. Thus, I refer here to the broader need for an improvement of communication, which would presumably guide the design of graduate programs.

First, the need for hydrologic scientists to stay sensitive to major socio-environmental questions requires that we take seriously the task of tracking and responding to the "web" of information, proposed earlier as a model for a hydrologic scientist's career. There is no substitute for broad and continuing education and flexibility of interest in nature and society. That point needs to be emphasized to graduate students, particularly Ph.D. candidates, who—contrary to many recent high-profile discussions of the future of graduate education—remain our best hope for the evolution of a creative hydrologic science contributing to human welfare. Although it is true that some mentors hire students as research assistants and treat them as drones and apprentices in the Dickensian way described in recent critiques of graduate education, it is not true that all Ph.D. students are obliged to focus narrowly on only their thesis topics, thus becoming "inflexible, illiterate, unemployable, etc." Nor is it true that they can only become informed about society's interests by taking social studies courses during their graduate careers. I encounter many students who entered graduate school with broad interests, an appetite for reading widely, learning languages, and other skills. Through preparing for careers in unpredictable futures, they learn a much wider range of skills than students of 20 years ago: they practice expressing themselves well in a variety of settings, they learn to teach well (or they have no chance of a tenure-track academic career these days), and they change their interests in response to emerging opportunities. Temporary intense focus on their thesis topics illustrates one of their strengths. However, it does not mean that they cannot stay informed about other subjects and practice the skills referred to above while they are working on their theses. Tracking and responding to the web of information, as referred to above, require mentoring through seminars, conversation, and the other intangible aspects of a lively intellectual climate that are recognizable in excellent graduate programs or, for that matter, in other well-led scientific agencies. Some suggestions for individual mentors, though not for leaders of entire programs, have been summarized in another report of the National Academy of Sciences (1997).

Second, since it has been emphasized by COHS and other bodies that hydrologic science is a geoscience, we need to develop some general knowledge of Earth as a system. This requires that we expand our horizons to stay informed about the other geosciences and the processes that they study. We also need to develop some appreciation of real geographical features, including why there are regional and secular differences in hydrologically significant characteristics of rainfall, soils, and vegetation. One can hardly expect hydrologists to develop fundamental yet realistic theories of hydrologic behavior unless they appreciate the range and pattern of the hydrologically significant properties of the planet and especially its continental surfaces.

Third, we need to develop a more coherent vocabulary, grounded in the philosophical principles of science. Mathematics is obviously a requirement because of the need to formalize theories for exact definition, communication, testing, and prediction. It is widely agreed that this is so, and to judge from journal articles, some progress has been made in improving the average analytical skill level of people entering the field of hydrology. However, Dooge, Klemes, and others have warned repeatedly that mathematical training alone is not sufficient to build hydrologic theory. Something more fundamental is required. We need to remind ourselves and our students of the principles and methods of science, including its goal of developing general theories of nature, its search for fundamental mechanisms, and its empirical tools both for exploration and hypothesis testing.

Lack of numeracy is declining as a communications block in the geosciences. Even many students who are esthetically attracted to traditionally descriptive field sciences are quite willing to study advanced mathematics, continuum mechanics, chemical kinetics, and other useful hydrologic tools given sufficient reason to do so. These are not the students who studied mathematics early in their careers because they were certain that it would be good for them. They still need convincing and motivating by concrete scientific applications before they will extend their mathematical skills. Helping students acquire these skills should be a goal of modern graduate education in hydrologic science. Hydrologic science is also recruiting students with strong analytical skills but little experience with landscapes, planets, processes, or measurement. It is difficult to expect students with no exposure to basic climatology, physical geography, or ecology to generate the idea of studying, say, the processes that lead to rainfall distributions over topography and their role in flood generation or how climatic changes might gradually affect plant community characteristics and hence the water balance. Thus, we also need to create some time in the education of these scientists to learn about the physics, chemistry, and biology of waters near the Earth's surface. New kinds of introductory graduate-level courses that present a quantitative and theoretical approach to some of these fields would probably attract both kinds of students described above and help them to recognize the overarching geoscience themes that proponents of hydrologic science have outlined.

At present, we do not communicate well enough to build hydrologic science into a broad, rapidly growing, socially useful activity. That problem can only be solved by behavioral changes forced by professors, hiring committees, employers, and editors. The crucial action eventually, however, will have to be curriculum reform. In a few universities that have the flexibility and funds to invest in new, sufficiently large Ph.D. programs in hydrologic science, it will be easier to hire and acculturate a diverse faculty covering the range of subjects referred to in the COHS report, provided that the leadership is open minded on this subject. At other universities a program in hydrologic science will have to be coordinated from offerings already in various departments of engineering, Earth science, atmospheric science, forestry, and related disciplines. The strains that Klemes

(1986) has described between hydrologic research and technological applications of hydrology in resource management are likely to be difficult to manage under such circumstances, unless the Earth or atmospheric science departments are stronger, larger, and more committed to studies of land surface processes than is usually the case. In the case of departmental or interdepartmental graduate programs in hydrologic science, both National Aeronautics Space Administration (NASA) Earth System Science fellowships and especially the original National Science Foundation (NSF) fellowships in hydrologic science (five per institution for five years) have been a crucial form of support that allows graduate students to pursue an interdisciplinary course between faculty advisers without having to devote a large amount of time to assisting the research of the faculty member.

Improving Measurement Capabilities

Measurement and even qualitative observation have always been weak components in the training of hydrologists. Most hydrology books never mention the characteristics of real land surfaces or describe processes occurring on them over a range of time scales. It is tacitly acknowledged that, because of the scale problem referred to in earlier sections, it is difficult to visually identify any characteristic or process being represented in a hydrologic model. Data, supplied from networks of rain gages or stream gages by the technical staffs of federal agencies, are usually taken at face value, their limitations given only passing mention. Most hydrology textbooks have early sections on how to fill gaps in data series and generally "make do" with whatever fragmentary data happen to be available in the project area of interest. This passivity has left most of us unskilled in conceiving of innovative and precise measurement techniques. Exceptions to this generalization are the skillful measurements of small-scale subsurface water storage and flow processes in some field and laboratory experiments and monitoring studies.

Yet there have recently been some technological revolutions affecting the availability, or in some cases the promise, of more and better data than have been available before. Despite recent retrenchments in some federal government stream-gaging programs, during the last decade of this century we are probably receiving more hydrologically relevant data than have been collected in the entire history of the science, and the pace of measurement shows no sign of slowing. Digital topography of the United States and most of the Earth's continents at spatial resolutions of 30 or 90 m is already, or soon will be, available, and higher-resolution data will emerge, as side-looking airborne radar is deployed from satellites and aircraft. Laser altimeters on aircraft already produce high-resolution topography for special purposes. Distributions of atmospheric water vapor are mapped with passive microwave sensors on spacecraft, and higher-resolution spatial and temporal distributions of individual rainstorms are measured with ground-level radar (Smith et al., 1996a, b). Global satellite measurements of radiation and surface temperature are available for monthly averages, with the promise of much higher resolution following the launch of satellites under the Earth Observing System and other programs by NASA and National Oceanic and Atmospheric Administration (NOAA) (Asrar and Dozier, 1994). Other data bases include regular measurements of snow distribution and the condition of plant covers, as well as low-frequency compilations of ground-level or satellite measurements of plant distributions, land cover, and soil properties. In the wings are promises of optical monitoring of snow cover (Rosenthal and Dozier, 1996), radar measurements of snow water equivalent (Shi and Dozier, 1996), high-precision topography for low-lying areas such as valley floors, and even surface soil moisture for some restricted range of environments. The entry of NASA and NOAA into the field of hydrology has thus been a revolutionary force, facilitating analyses that were simply impossible earlier. In the United States, although there has been some reduction in the number of monitoring stations, even traditional measurements of rainfall, streamflow, water chemistry, and soil properties are more easily available than ever because of a vast effort by some federal agencies to disseminate data through electronic media.

These data sources are not without blemish, of course. The resolution of routinely available digital topography is still too coarse to reflect the scale of the dynamics of runoff and erosion; radar rainfall is difficult to calibrate and interpret; measurements of plant conditions are compromised by atmospheric aerosols and other effects; and the litany of difficulties goes on. Many uncertainties remain in interpreting the ground-level radiation signal received at the top of the atmosphere. However, all of the examples represent major improvements in useful hydrologic data, especially in terms of spatial coverage. They are simply too promising and pervasive to be ignored in the training and retraining of hydrologic scientists. A significant time commitment to remote sensing and spatial data handling is now required in hydrologic training, including time spent critically reviewing the relationship between the interpreted product and actual ground conditions (which themselves may be difficult to define). Remotely sensed products allow us to observe large remote features such as entire continents, river basins, mountain ranges, and floodplains. Thus, they are crucial to our ability to observe large-scale processes such as floodplain inundation (Sippel et al., 1994; Vorosmarty et al., 1996; Mertes, 1997), rainfall fields (Smith, 1996b), and the generation of massive sediment pulses from regional-scale intense rainfall. They also contribute to the goal of constructing credible spatially distributed runoff and evaporation models.

Most of the new data sources referred to above are being delivered to hydrology without being ordered. In the traditional manner of "making do" with data from networks installed for purposes associated with water management, surface water hydrologists are muddling through, grateful for every new bit of data we can get our hands on. In particular, the satellite-based sensors were developed for other reasons and have been turned to purposes useful to us mainly by geophysicists, atmospheric scientists, and others. Some of these groups are now in a position to recommend satellite and other remote sensing programs to provide critical measurements for their scientific purposes. However, among hydrologic scientists, only hydrometeorologists seem in a position to make such requests, because of the previously mentioned lack of theoretical convergence and agenda building among hydrologists. The chance to guide hydrologic data collection could be an important product of agenda building.

A particular example of directed data collection that was highlighted by COHS was the opportunity to participate in large, coordinated, multiinvestigator field campaigns, such as the First ISLSCP (International Satellite Land Surface Climatology Program) Field Experiment (FIFE), Hydrologic-Atmospheric Pilot Experiments (HAPEX) and Boreal Ecosystem-Atmosphere Study (BOREAS), and the upcoming Large-Scale Biosphere-Atmosphere in Amazonia (LBA) (Amazon Basin) and GEWEX (Global Energy and Water Cycle Experiment) Continental-Scale International Project (GCIP) (mainly Mississippi River basin) activities. So far, it is the hydrometeorologists who have been able to take advantage of these opportunities, and surface water hydrologists continue to lag behind in the effectiveness of their data requests.

Oversight

Though referred to above as a fledgling, hydrologic science is important to transcendent societal concerns such as the reciprocal interaction between humans and climate, global sustainability, environmental justice between nations and generations, and the influence of continental perturbations on the nearshore ("green-water") ocean. For this reason the science needs to be fostered as a strategic concern—perhaps as the special concern of some standing oversight body, analogous to the Climate Research Committee of the National Research Council (NRC), a committee overseen by the Board on Atmospheric Sciences and Climate. The most appropriate venue for such a Water Science Committee could be the NRC's Water Science and Technology Board (WSTB). The WSTB already works for hydrologic science, including its sponsorship of the original Committee on Opportunities in the Hydrologic Sciences and in fact has been the only such voice in this country in recent decades. However, most of the board's work is ultimately related to the technological issues that reflect the interests of the federal agencies that support the WSTB. Creating a Water Science Committee to nurture and develop hydrologic science would provide for a separation of emphases similar to that being proposed among hydrologists themselves.

It would be desirable for such a committee to continue fostering hydrologic science, more or less as described in the original COHS report but with more direct acknowledgment of the ethical responsibilities of environmental scientists to work on problems of broad social concern. The standing committee would be composed of a small, broadly informed group of hydrologists with a range of backgrounds who take seriously the responsibility for hydrologic science as a whole, rather than representing, say geomorphology, ground water hydrology, or hydrometeorology. They would have to see themselves as trustees, representing the interests of the next generation of hydrologists and all the nonhydrologists who pay the bills of the science.

The committee members could articulate trends or gaps in knowledge, thereby providing continuous advice to the directors of funding programs and building consensus about research strategies for the science. The committee could sponsor regional seminars and special sessions at professional meetings. It could publish occasional commentaries. Dissemination of information about the contributions of hydrologic science to human knowledge and its plans for extending that knowledge would be another important activity. Several other sciences, most notably astronomy, have proven that frequent representation on the weekly science pages of the *New York Times*, combined with a strong community research agenda skillfully communicated to Congress, seem to be associated with the ability to mobilize massive investment in that research agenda. On the water planet, hydrologic science should be at least as diverting as the stars.

In its oversight of the science the committee could take some responsibility for maximizing the nation's entire investment in hydrologic research by promoting interactions between academe and the federal agencies interested in water. This interaction has declined precipitously during the past 10 to 15 years as federal budgets have tightened. However, communication depends on attitudes, even in the absence of money. A decline in optimism has also reduced the probability of new research collaborations and the transmission of ideas. During times of slow hiring, federal agencies become isolated from the stream of bright young people who continue to pass through universities. Academics lose contact with valued colleagues, underutilized data sources and equipment, and interesting scientific problems motivated by the agencies' responsibilities. This running down of the federal-academe relationship is a loss to the nation, and it needs to be reversed at little or no cost. Exchanging information and sharing joint responsibility for a national scientific committee with an optimistic charge could initiate this reversal. Visits and expressions of interest in each other's programs (research and nonresearch) would be easy steps in reestablishing a fruitful relationship.

Concerning how to fund such an activity, a straw proposal is presented to provoke thought among those who look wearily on any proposal for new activity at this time. A small and active (publishing several essay-length reports per year) standing NRC committee might cost $150,000 per year. Six annual contributions of $25,000 would suffice. These contributions could be sought among a variety of institutions that stand to gain from direct representation on the committee or from a continuous stream of ideas about opportunities for research that might be useful to its operation. The directors of the NSF and NASA hydrology programs might find that such information would reduce their need for outside advice from current sources, so that they might be able to divert some of the resources they currently use for that purpose. Some federal agencies with a need to keep informed about developments in scientific hydrology might also be induced to contribute. Since secondary goals of the committee would be to advertise widely the contributions of scientific hydrology and to stimulate agency-academe productivity, there are reasons for the committee to be useful to several agencies. Finally, the national organizations that represent private interests in water and power might also be induced to provide stable annual funding for a committee with a long-range strategic view of the science.

SUMMARY

This author does not share the Sputnik-era confidence of some colleagues that hydrologic science left to its own devices would automatically improve human welfare. However, a modestly supervised hydrologic science, imbued with strong philosophical and ethical principles about the conduct of scientific research on behalf of society, could be of enormous benefit to the nation and to its collateral field of applied hydrology. Much good hydrologic science can already be found in the premier journals of hydrology, but it is spread so thinly amid excellent representations of other types of work that the ethos of hydrologic science does not emerge. As Klemes (1988) has pointed out, the science and the nonscience in hydrology frequently become mixed up, to the confusion of both.

Before a strong hydrologic science can grow and interact productively with the other geosciences, some actions and permanent behavioral changes are needed. We need to develop more unified approaches in our choice of important research targets and in our quest for theoretical generalizations about fundamental processes. We must put aside differences of background and have the patience to communicate with each other so that everyone understands the current state of knowledge, or at least knows exactly why he or she disagrees with it. We have to extend our ability to use or at least to understand a wide variety of new technologies that for the first time offer to measure the spatial characteristics of hydrologic processes and characteristics at scales up to regional and global. Finally, the oversight mentioned above needs to be provided by the NRC, which would not only act as an authoritative voice on scientific hydrology but also generate a stream of creative advice about continuing opportunities in hydrologic science.

ACKNOWLEDGMENTS

I am very grateful to Steve Burges and Jeff Dozier, who in many conversations helped to clarify some of the issues raised here, and to Bill Dietrich and Laura Ehlers for reviewing the manuscript.

REFERENCES

Abrahams, A. D., and J. J. Ponczynski. 1976. Drainage density in relation to precipitation intensity in the U.S.A. J. Hydrol. 75:383-388.

Ad Hoc Panel on Hydrology. 1962. Scientific Hydrology. U.S. Federal Council for Science and Technology, Washington, D.C.

Ahnert, F. 1976. Brief description of a comprehensive three-dimensional process-response model of landform development. Z. Geomorphol. 25(Suppl.):29-49.

Asrar, G., and J. Dozier. 1994. Science Strategy for the Earth Observing System. Woodbury, N.Y.: American Institute of Physics Press.

Back, W. 1991. Opportunities in the hydrological sciences. EOS Am. Geophys. Union Trans. 72:491-492.

Baker, V. R. 1993. Flood hazards–learning from the past. Nature 361(6411):402-403.

Benda, L. E., and T. Dunne. 1998a. Stochastic forcing of sediment supply to channel networks by landsliding and debris flow. Water Resour. Res. 33:2849-2863.

Benda, L. E., and T. Dunne. 1998b. Stochastic forcing of sediment transport in channel networks. Water Resour. Res. 33:2865-2880.

Benda, L. E., D. J. Miller, T. Dunne, G. H. Reeves, and J. K. Agee. In press. Dynamic landscape systems. In Ecology and Management of Streams and Rivers in the Pacific Northwest Coastal Ecoregion. R. J. Naiman and R. E. Bilby, eds. New York: Springer-Verlag.

Beven, K. J. 1986. Hillslope runoff processes and flood frequency characteristics. Pp. 187-202 in Hillslope Processes. A. D. Abrahams, ed. St. Leonards, Australia: Allen & Unwin.

Beven, K. 1987. Towards a new paradigm in hydrology. In Proc. Symp. On Water for the Future Wallingtonford, U.K.: International Association of Hydrological Sciences.

Beven, K. 1989. Changing ideas in hydrology—the case of physically-based models. J. Hydrol. 105:157-172.

Beven, K. J. 1993. Prophecy, reality, and uncertainty in distributed hydrologic modeling. Adv. Water Resour. 16:41-51.

Beven, K. 1996. Equifinality and uncertainty in geomorphological modeling. Pp. 289-313 in The Scientific Nature of Geomorphology. B. L. Rhoads and C. E. Thorn, eds. New York: John Wiley & Sons.

Beven, K. J., and A. M. Binley. 1992. The future of distributed models: Model calibration and uncertainty prediction. Hydrol. Process. 6:279-298.

Beven, K., and M. Kirkby. 1979. A physically-based variable contributing area model of basin hydrology. Hydrol. Sci. Bull. 24:43-69.

Beven, K. J., E. F. Wood, and M. Sivapalan. 1988. On hydrological heterogeneity—catchment morphology and catchment response. J. Hydrol. 100:353-375.

Black, P. E. 1995. The critical role of "unused" resources. Water Resour. Bull. 31:589-592.

Blöschl, G., and M. Sivapalan. 1995. Scale issues in hydrologic modeling: A review. Hydrol. Process. 9:251-290.

Bush, V. 1945. Science—The Endless Frontier. Washington, D.C.: National Research Council.

Calder, I. R. 1990. Evaporation in the Uplands. Chichester, U.K.: Wiley.

Calder, I. R., and M. D. Newson. 1979. Land-use and upland water resources in Britain—a strategic look. Water Resour. Bull. 15:1628-1639.

Cayan, D. R., and D. H. Peterson. 1990. The influence of North Pacific circulation on streamflow in the west. Pp. 375-398 in Aspects of Climate Variability in the Pacific and Western Americas. D. H. Peterson, ed. Washington, D.C.: American Geophysical Union.

Chagnon, S. A. 1985. Research agenda for floods to solve a policy failure. Am. Soc. Civ. Eng. J. Water Resour. Plan. Manage. 111:54-64.

Crozier, M. J., E. E. Vaughan, and J. M. Tippett. 1990. Relative instability of colluvium-filled bedrock depressions. Earth Surf. Process. Landforms 15(4):329-339.

Dietrich, W. E., and T. Dunne. 1978. Sediment budget for a small catchment in mountainous terrain. Z. Geomorphol. 29(Suppl.):191-206.

Dietrich, W. E., and T. Dunne. 1993. The channel head. Pp. 175-220 in Channel Networks: A Geomorphological Perspective. K. J. Beven and M. J. Kirkby, eds. Chichester, U.K.: John Wiley & Sons.

Dietrich, W. E., R. Reiss, M-L. Hsu, and D. R. Montgomery. 1995. A process-based model for colluvial soil depth and shallow landsliding using digital elevation data. Hydrol. Process. 9:383-400.

Dooge, J. C. 1986. Looking for hydrologic laws. Water Resour. Res. 22:46S-58S.

Dooge, J. C. 1988. Hydrology in perspective. Hydrol. Sci. J. 33:61-85.

Duffy, C. J. 1988. Groundwater circulation in a closed desert basin: Topographic scaling and climatic forcing. Water Resour. Res. 24:1675-1688.

Dunne, T. 1980. Formation and controls of channel networks. Prog. Phys. Geog. 4:213-239.

Dunne, T., T. R. Moore, and C. H. Taylor. 1975. Recognition and prediction of runoff producing zones in humid areas. Hydrol. Sci. Bull. 20:305-327.

Dunne, T., L. A. K. Mertes, R. H. Meade, J. E. Richey, and B. R. Forsberg. 1998 Exchanges of sediment between the floodplain and channel of the Amazon River in Brazil. Geol. Soc. Am. Bull. 110:450-467.

Eagleson, P. S. 1978. Climate, soil, and vegetation. Water Resour. Res. 15:705-776.

Gash, J. H. C., C. A. Nobre, J. M. Roberts, and R. L. Victoria (eds.). 1996. Amazonian Deforestation and Climate. Chichester, U.K.: John Wiley & Sons.

Gell-Mann, M. 1994. The Quark and the Jaguar: Adventures in the Simple and the Complex. New York: W. H. Freeman.

Gomez, B., L. A. K. Mertes, J. D. Phillips, F. J. Magilligan, and L. A. James. 1995. Sediment Characteristics of an Extreme Flood: 1993 Upper Mississippi River Valley. Geology 23:963-966.

Graf, W. L. 1994. Plutonium in the Rio Grande: Environmental Change and Contamination in the Nuclear Age. New York: Oxford University Press.

Grayson, R. B., I. D. Moore, and T. A. McMahon. 1992. Physically based hydrologic modeling. 2. Is the concept realistic? Water Resour. Res. 28:2659-2666.

Gupta, V. K., E. Waymire, and C. T. Wang. 1980. A representation of an instantaneous unit hydrograph from geomorphology. Water Resour. Res. 16:855-862.

Horton, R. E. 1945. Erosional development of streams and their drainage basins: Hydrophysical approach to quantitative morphology. Geol. Soc. Am. Bull. 56:275-370.

Jakeman, A. J., and G. M. Hornberger. 1993. How much complexity is warranted in a rainfall-runoff model? Water Resour. Res. 29:2637-2649.

Jones, J. A., and G. E. Grant. 1996. Peak flow responses to clear-cutting and roads in small and large basins, western Cascades, Oregon. Water Resour. Res. 32:959-974.

Junk, W., P. B. Bayley, and R. E. Sparks. 1989. The flood-pulse concept in river-floodplain systems. Pp. 110-127 in Proceedings of the Large River Symposium. D. P. Doge, ed. Otttawa: Canadian Department of Fisheries and Oceans.

Kesel, R. H., E. G. Yodis, and D. J. McCraw. 1992. An approximation of the sediment budget of the lower Mississippi River prior to major human modification. Earth Surface Process. Landforms 17:711-722.

Kirkby, M. J. 1986. A two-dimensional simulation model for slope and stream evolution. Pp. 203-222 in Hillslope Processes. A. D. Abrahams, ed. St. Leonards, Australia: Allen and Unwin.

Kirkby, M. J., and R. J. Chorley. 1967. Throughflow, overland flow and erosion. Bull. Int. Assoc. Sci. Hydrol. 12:5-21.

Klemes, V. 1982. Empirical and causal models in hydrology. Pp. 95-104 in Scientific Basis of Water Resource Management. Washington, D.C.: National Academy Press.

Klemes, V. 1986. Dilettantism in hydrology: Transition or destiny? Water Resour. Res. 22:177S-188S.

Klemes, V. 1988. A hydrological perspective. J. Hydrol. 100:3-28.

Klemes, V. 1989. The improbable probabilities of extreme floods and droughts. Pp. 43-51 in Hydrology and Disasters. O. Starosolszky and O. M. Melder, eds. London: James and James.

Law, F. 1956. The effect of afforestation on the yield of water catchment areas. J. Br. Waterwks. Assoc. 38:489-494.

Law, F. 1957. Measurement of rainfall, interception, and evaporation losses in a plantation of Sitka spruce trees. Int. Assoc. Hydrol. Sci. 44:397-411.

Lewin, J., B. E. Davies, and P. J. Wolfenden. 1977. Interactions between channel change and historic mining sediments. Pp. 353-368 in River Channel Changes. K. J. Gregory, ed. Chichester, U.K.: John Wiley.

Lettenmaier, D. P., and T. Y. Gan. 1990. Hydrologic sensitivities of the Sacramento-San Joaquin River basin, California, to global warming. Water Resour. Res. 26:69-86.

Loague, K. 1990. R-5 revisited 2. Reevaluation of a quasi-physically based rainfall-runoff model with supplemental information. Water Resour. Res. 26:973-987.

Maidment, D. R. (ed.). 1993. Handbook of Hydrology. New York: McGraw-Hill.

Marengo, J. A., J. R. Miller, G. L. Russell, C. E. Rosenzweig, and F. Abramopoulos. 1994. Calculations of river-runoff in the GISS GCM: Impact of a new land-surface parameterization and runoff routing model on the hydrology of the Amazon River. Clim. Dynam. 10:349-361.

Marron, D. C. 1992. Floodplain storage of mine tailings in the Belle Fourche river system: A sediment budget approach. Earth Sur. Process. Landforms 17:675-685.

Meade, R. H. (ed.). 1995. Contaminants in the Mississippi River, 1987-92. U.S. Geol. Surv. Circ. 1133. Denver, Colorado, U.S. Geological Survey.

Meade, R. H., T. R. Yuzyk, and T. J. Day. 1990. Movement and storage of sediment in rivers of the United States and Canada. Pp. 255-280 in The Geology of North America, O-1: Surface Water Hydrology. M. G. Wolman and H. C. Riggs, eds. Boulder: Geological Society of America.

Meinzer, O. E. 1942. Hydrology. New York: Dover Publications.

Mertes, L. A. K. 1997. Description and significance of the perirheic zone on inundated floodplains. Water Resour. Res. 33:1749-1762.

Mertes, L. A. K., T. Dunne, and L. A. Martinelli. 1996. Channel-floodplain geomorphology along the Solimões-Amazon River, Brazil. Geol. Soc. Am. Bull. 108:1089-1107.

Meyboom, P. 1962. Patterns of groundwater flow in the prairie profile. Pp. 5-20 in Proc. Hydrology Symposium No. 3. Ottawa, Ontario: National Research Council of Canada.

Miller, N. L., and J. Kim. 1996. Numerical prediction of precipitation and river flow over the Russian River watershed during the January 1995 California storms. Bull. Am. Meteorol. Soc. 77:101-105.

Moore, I. D., P. E. Gessler, G. A. Nielsen, and G. A. Peterson. 1993. Soil attribute prediction using terrain analysis. Soil Sci. Soc. Am. J. 57:443-452.

Myers, M. F., and G. F. White. 1993. The challenge of the Mississippi flood. Environment 35:6-35.

Nash, L. L., and P. H. Gleick. 1991. Sensitivity of streamflow in the Colorado basin to climatic changes. J. Hydrol. 125:221-241.

Nash, J. E., P. S. Eagleson, J. R. Philip, and W. H. Van der Molen. 1990. The education of hydrologists. Hydrol. Sci. J. 35:597-607.

National Academy of Sciences. 1997. Advisor, Teacher, Role Model, Friend: On Being a Mentor to Students in Science and Engineering. Washington, D.C.: National Academy Press.

National Research Council. 1991. Opportunities in the Hydrologic Sciences. Washington, D.C.: National Academy Press.

National Research Council. 1995. A Review of the U.S. Global Change Research Program and NASA's Mission to Planet Earth/Earth Observing System. Washington, D.C.: National Academy Press.

Nicholas, A. P., and D. E. Walling. 1997. Modeling Flood Hydraulics and Overbank Deposition on River Floodplains. Earth Sur. Process. and Landforms 22(N1):59-77.

Petroski, H. 1997. Development and Research. Am. Sci. 85:210-213.

Potter, P. E. 1978. The significance and origin of big rivers. J. Geol. 86:13-33.

Power, M. E., A. Sun, G. Parker, W. E. Dietrich, and J. T. Wooton. 1995. Hydraulic food-chain models. BioScience 45:159-167.

Redmond, K. T., and R. W. Koch. 1991. Surface climate and streamflow variability in the western United States and their relationship to large-scale circulation indices. Water Resour. Res. 27:2381-2399.

Robinson, J. S., M. Sivapalan, and J. D. Snell. 1995. On the relative roles of hillslope processes, channel routing, and network geomorphology in the hydrologic response of natural catchments. Water Resour. Res. 31:3089-3101.

Rodríguez-Iturbe, I., and J. B. Valdes. 1979. The geomorphological structure of hydrologic response. Water Resour. Res. 15(6):1435-1444.

Rosenthal, W., and J. Dozier. 1996. Automated mapping of montane snow cover at subpixel resolution from the Landsat Thematic Mapper. Water Resour. Res. 32:115-130.

Rutter, A. J. 1967. An analysis of evaporation from a stand of Scots pine. Pp. 403-418 in Forest Hydrology. W. E. Sopper and H. W. Lull, eds. Oxford, England: Pergamon Press.

Sellers, P. J., R. E. Dickinson, D. A. Randall, A. K. Betts, F. G. Hall, J. A. Berry, G. J. Collatz, A. S. Denning, H. A. Mooney, C. A. Nobre, N. Sato, C. B. Field, and A. Henderson-Sellers. 1997. Modeling the exchanges of energy, water, and carbon between continents and the atmosphere. Science 275:502-505.

Seyfried, M. S., and B. P. Wilcox. 1995. Scale and the nature of spatial variability: Field examples having implications for hydrologic modeling. Water Resour. Res. 31:173-184.

Shi, J., and J. Dozier. 1996. Estimation of snow water equivalence using SIR-C/X-SAR. 1996 IEEE Proceedings from International Geoscience and Remote Sensing Symposium, 96CH35875IV:2002-2004. Piscataway, New Jersey: Institute of Electrical and Electronics Engineers.

Shuttleworth, W. J. 1988. Macrohydrology: The new challenge for process hydrology. J. Hydrol. 100:31-56.

Shuttleworth, W. J. 1989. Micrometeorology of temperate and tropical forest. Philos. Trans. R. Soc. London, Ser. B. 324:299-334.

Shuttleworth, W. J. 1991. Evaporation models in hydrology. Pp. 93-120 in Land Surface Evaporation: Measurement and Parameterization. T. J. Schmugge and J-C. André, eds. New York: Springer Verlag.

Shuttleworth, W. J., J. H. C. Gash, C. R. Lloyd, C. J. Moore, J. Roberts, A. O. de Marques, G. Fisch, V. P. de Silva, M. N. Ribeiro, L. C. B. Molion, L. D. A. de Sa, J. C. Nobre, O. M. R. Cabral, S. R. Patel, and J. C. Moraes. 1984. Eddy correlation measurements of energy partition for Amazonian forest. Q. J. R. Meteorol. Soc. 110:1143-1162.

Sippel, S. K., S. K. Hamilton, J. M. Melack, and B. Choudhury. 1994. Passive microwave satellite observations of seasonal variations of inundation area in the Amazon River floodplain, Brazil. Remote Sensing Environ. 4:70-76.

Smith, J. A., D. J. Seko, M. L. Baeck, and M. D. Hurlow. 1996a. An intercomparison study of NEXRAD precipitation studies. Water Resour. Res. 32:2035-2045.

Smith, J. A., M. L. Baeck, M. Steiner, and A. J. Miller. 1996b. Catastrophic rainfall from an upslope thunderstorm in the central Appalachians: The Rapidan storm of June 27, 1995. Water Resour. Res. 32:3099-3113.

Smith, T. R., B. Birnir, and G. E. Merchant. 1998a. Towards an elementary theory of drainage basin evolution: I. The theoretical basis. Compu. and Geosci. 23(9): 811-822.

Smith, T. R., G. E. Merchant, and B. Birnir. 1998b. Towards an elementary theory of drainage basin evolution: II. A computational evaluation. Compu. and Geosci. 23(9): 823-849.

Stanford, J. A., and J. V. Ward. 1993. An ecosystem perspective of alluvial rivers: Connectivity and the hyporheic corridor. J. N. Am. Benthol. Soc. 12:48-60.

Stedinger, J. R., and V. R. Baker. 1987. Surface water hydrology: Historical and paleoflood information. Rev. Geophys. 25:119-124.

Toth, J. 1963. A theoretical analysis of ground water flow in small drainage basins. J. Geophys. Res. 68:4795-4812.

Toth, J. 1966. Mapping and interpretation of field phenomena for groundwater reconnaissance in a Prairie environment, Alberta, Canada. Bull. Int. Assoc. Sci. Hydrol. 11(2):20-68.

Trimble, S. W. 1983. A sediment budget for Coon Creek basin in the driftless area, Wisconsin, 1853-1977. Am. J. Sci. 283:454-474.

Vannote, R. L., G. W. Minshall, K. W. Cummins, J. R. Sedell, and C. E. Cushing. 1980. The river continuum concept. Can. J. Fish. Aquat. Sci. 37:130-137.

Vorosmarty, C. J., C. J. Willmott, B. J. Choudhury, A. L. Schloss, T. K. Stearns, S. M. Robeson, and T. J. Dorman. 1996. Analyzing the discharge regime of a large tropical river through remote sensing, ground-based climatic data, and modeling. Water Resour. Res. 32:3137-3150.

Webb, R. H., and J. L. Betancourt. 1990. Climatic effects on flood frequency: An example from southern Arizona. Pp. 61-66 in Proceedings of the. Sixth Annual Pacific Climate (PACLIM) Workshop. J. L. Betancourt and A. M. Mackay, eds. California Department of Water Resources, Sacramento: Interagency Ecological Studies Program for the Sacramento-San Joaquin Estuary, Technical Rep. 23.

Weinberg, A. M. 1972. Science and trans-science. Pp. 105-122 in Civilization and Science: In Conflict or Collaboration? Amsterdam: Elsevier.

Whalton, P. H., D. Gilman, and M. A. J. Williams. 1990. Rainfall and river flow variability in Asia, Australia, and East Africa linked to El Niño—Southern Oscillation events. Pp. 71-82 in Lessons for Human Survival: Nature's Record from the Quaternary. P. Bishop, ed. Sydney: Geological Society of Australia.

Wigley, T. M. L., and P. D. Jones. 1985. Influence of precipitation changes and direct CO_2 effects on streamflow. Nature 314:149-151.

Willgoose, G. R., R. L. Bras, and I. Rodriguez-Iturbe. 1991. A physically-based coupled network growth and hillslope evolution model. 1. Theory. Water Resour. Res. 27:1671-1684.

Wolock, D. M., and G. M. Hornberger. 1991. Hydrological effects of changes in levels of atmospheric carbon dioxide. J. Forecast. 10:105-116.

Wood, E. F. 1995. Scaling behavior of hydrological fluxes and variables: Empirical studies using a hydrological model and remote sensing data. Hydrol. Process. 10:21-36.

Wood, E. F., M. Sivapalan, K. Beven, and L. Band. 1988. Effects of spatial variability and scale with implications to hydrologic modeling. J. Hydrol. 102:29-47.

Ziemer, R. R., and J. S. Albright. 1987. Subsurface pipeflow dynamics of north-coastal California swale systems. Pp. 71-90 in Erosion and Sedimentation in the Pacific Rim. R. Beschta, T. Blinn, G. E. Grant, F. J. Swanson and G. G. Ice, eds. Wallingford, United Kingdom: International Association of Hydrological Sciences.

2

Aquatic Ecosystems: Defined by Hydrology.
Holistic Approaches Required for Understanding, Utilizing, and Protecting Freshwater Resources

Diane M. McKnight
Institute of Arctic and Alpine Research
University of Colorado, Boulder

ABSTRACT

The scientific disciplines of hydrology and limnology are distinct though closely connected. Limnology is an integrative discipline, applying physics, chemistry, and biology to the study of inland aquatic ecosystems. Inland aquatic ecosystems include streams, rivers, lakes, reservoirs, and wetlands and possess many diverse characteristics. The focus of this paper is how hydrology defines aquatic ecosystems, especially the ecosystem boundaries and the fluxes of water, solutes, organisms, and detrital organic matter across boundaries. For millennia, human civilizations have used knowledge of hydraulics and hydrology to distribute water resources for agricultural, municipal, recreational, and power generation purposes. In contrast, the basic concepts of ecology have been established only during this century, and their application to the management of water resources, through watershed management or ecosystem management approaches, is just now coming into practice. Undoubtedly, freshwater is a strategic resource and advances in hydrology, limnology, and water resource research and management will be required in the future. Broader education of all scientists and engineers involved in the study and management of freshwater is one way to promote greater knowledge and expertise and to achieve more complete, holistic resolutions of water resource issues. Further, it is critical to maintain and expand networks of monitoring programs that provide long-term records of physical and chemical characteristics of inland aquatic ecosystems. These records allow us to extend current understanding of aquatic ecosystems to the appropriate hydrologic time scales and will allow for resolution of water resource issues through a process of "knowledge-based" consensus at local and regional scales.

INTRODUCTION

Freshwater is a strategic resource upon which human society depends. Water is consumed for drinking water, industrial uses, and irrigation, and flow regulation is instrumental to power generation and flood protection. Aquatic ecosystems provide fish and other food resources as well as recreational resources, which are highly valued by communities. All of these uses depend not only upon having a sufficient quantity of water but also on the quality of water, that is, the physical and chemical characteristics of water. As we gain a greater scientific understanding of aquatic ecosystems, it is apparent that the quality of aquatic ecosystems influences not only the resident aquatic biota but also the quality of a water body as a water resource. These interactions can be represented as healthy aquatic ecosystems providing "goods and services" to the other users of water resources, as shown in Figure 1.

In many areas of the world and in some regions of the United States, water is now a limited strategic resource. Over the next 25 to 50 years, constraints on the

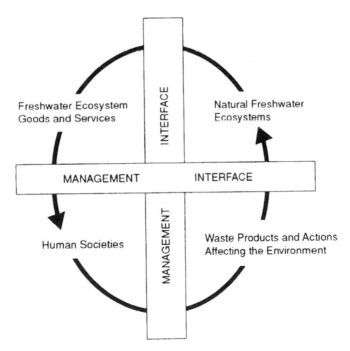

FIGURE 1   Schematic diagram illustrating the provision of goods and services by aquatic ecosystems and the role of the management interface in these interactions. Source: Reprinted, with permission, from Naiman et al. (1995). © 1995 from Island Press.

expansion of human societies and on economic development associated with limitations in water quantity and quality are expected to become more severe. There will not be magic bullets, quick fixes, or easy answers to the water resource challenges that lie ahead. Restructuring, optimization, and otherwise fine-tuning of existing water resource systems to achieve multiple goals will be required, as well as increasing the efficiency of irrigation and other consumptive uses (Postel, 1997). In the United States, federal, state, and local governments provide an infrastructure to put into place changes at a scale commensurate with that of large river basins. Some steps have been made in this direction, such as the recent landmark agreement that regulates flow in the Colorado River to achieve power generation and ecological goals. The governmental infrastructure and the commitment to multiple goals are important assets but are, in and of themselves, insufficient to meet the water resource challenges of the future. Clearly, these assets must be matched with greater knowledge in all aspects of aquatic sciences and engineering.

We need not only more in-depth understanding of hydrologic, chemical, biological, and ecological processes but also more detailed characterization of local and regional hydrology and aquatic ecology. Both predictive understanding of processes and knowledge of detailed characteristics are needed for effective fine-tuning at the scale of an individual reservoir, river, lake, wetland, or ground water aquifer. Management of water resources has been and will continue to be a governmental function. In some regions, private nongovernmental bodies, such as the Hudson River Foundation, also are playing an active role in research and management. Given that greater scientific knowledge will continue to promote progress in water resources and other civic issues, we can make the case that now is the time for federal, state, and local governments to invest in aquatic science. In response to the argument that a greater investment in research and training in aquatic science is not affordable within the current constrained governmental budgets, it can only be pointed out that the consequences of lack of knowledge and scientifically ill-founded or unsustainable practices are potentially disastrous as demands for water resources continue to grow. Simply put, greater knowledge of aquatic science will provide for the future, while lack of knowledge may have dire consequences.

Together, the disciplines of hydrology and limnology comprise a healthy portion of aquatic science. In this paper, hydrology is intended to include the study of the hydrologic cycle and physical processes controlling the movement of water in surface and ground waters. Limnology is broadly defined as an integrative discipline applying physics, chemistry, and biology to study inland aquatic ecosystems, which include streams, rivers, lakes, reservoirs, and wetlands. In order to meet the water resource challenges of the future, investments should be made now to advance these two disciplines and to provide comprehensive training of scientists and engineers in these disciplines.

Hydrology and limnology are intimately related. Despite their different histories and etiologies, there are some common approaches that can facilitate interaction and interdisciplinary research. In fact, the most exciting areas of research currently include those based upon integration of hydrologic and ecological approaches. This paper will conclude with recommendations for revitalizing education in limnology to promote greater knowledge of hydrologic processes among future limnologists. Similar recommendations are presented for incorporating limnology and ecology into the training of future hydrologists and water resource engineers.

Different Histories of Hydrology and Limnology

For millennia the progress of human societies has been tied to the development and use of water resources. Developing cities and agrarian communities put to good use the basic idea that water flows downhill, rainfall to river to ocean. Engineering know-how preceded the quantitative sciences of hydraulics and hydrology by many centuries. The skills to route water through channels, aqueducts, and pipes were well developed in Roman times. The plumbing systems of the Roman baths are masterpieces of civil engineering, one of which can be observed in an operational state in the museum in Bath, England (Cunliffe, 1993). The large reservoir, sluices, and lead pipe drain were constructed on the frontier of the Roman Empire beginning in the first century A.D., 1,500 years before the equations governing flow in pipes were developed by Bernoulli. In the new world, the Anasazi, the ancient Pueblo people of the southwest, moved water by sluices and man-made canals around 1000 A.D. Thus, hydraulics and hydrology were derived from physics and geology but were developed with an extensive foundation of empirical knowledge from observation and engineering experience. Further, new conceptual advances were rapidly employed as water resource systems were developed and refined.

The most remarkable changes to aquatic ecosystems, as a result of human activity, came with the development of agriculture. These alterations have accelerated in recent centuries and were generally not intended to improve water quality or the health of an aquatic ecosystem. In this century the motivation for extensively studying aquatic ecosystems stems from problems associated with this inadvertent degradation. Issues that have received considerable attention over the past 50 years include the discharge of nutrients from wastewater treatment plants, leading to eutrophication, and the atmospheric release of sulfur and nitrogen oxides during fossil fuel combustion, leading to acid rain and acidification of water bodies far removed from the sources of pollution.

In contrast to hydrology, limnology is based upon a concept that is less than a century old. In 1991 Science (Pool, 1991) listed the top 20 most significant scientific concepts of our century. The central theme of ecology, that "all life is connected," was listed with three other biological concepts (all organisms are made of cells, life is based on a genetic code, life evolves through natural selection). Continental drift, during which the Earth's surface is in continual change,

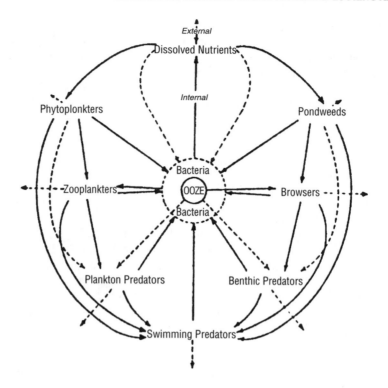

FIGURE 2   Lindeman's diagram of a food cycle in Cedar Bog Lake.   Source: Reprinted, with permission, from Lindeman (1941). © by the American Midland Naturalist.

was also identified as a new significant concept. Together these concepts have radically changed how we view and understand the natural world. The term ecosystem was introduced by Alfred George Tansley in 1935 and was presented as a basic unit of nature in the continuum from the atom to the universe (Golley, 1993). Although the ecosystem concept is to some extent based upon the descriptive information provided by naturalists and on the work of community ecologists, it represents an integration of biology with the more quantitative sciences of physics and chemistry (Mayr, 1982).

Although limnology is often viewed as a subset of ecology, the two fields developed in parallel (Golley, 1993). While the ecosystem concept is credited to Tansley, limnologists had previously voiced similar ideas that coupled the biota to their physical environment. Forbes' description of "the lake as a microcosm" in 1892 was further developed by the work of Birge and Juday. Lindeman was the first to explicitly apply Tansley's ecosystem concept in his classic work on Cedar Bog Lake. From his study of a small lake, he concluded that the ecosystem was the fundamental unit of trophic-dynamics, during which energy is transferred from primary producers (photosynthesizing algae and plants) to heterotrophs (grazers and predators). Figure 2 presents a diagram from his paper in *American Midland Naturalist* illustrating this concept.

The next major advance in limnology was the application of the small watershed approach to quantitatively understand the relationships between biological, chemical, and hydrologic processes in a watershed. This work began in 1963 with the Hubbard Brook Ecosystem Study, a collaborative effort between ecologists, hydrologists, and geochemists. The study quantified the input and output of chemicals to and from the watershed in constructing a nutrient budget for a forest ecosystem in New England (Likens et al., 1977). It has served as a basis for understanding management practices for New England forests and the effects of acid rain in the region. The streamflow and water quality records for Hubbard Brook are the primary data sets for this ongoing study.

In 1980 Vannote et al. presented a theoretical concept describing how ecological characteristics would vary along a "continuum" from a headwater stream to an intermediate-order stream to a large river. This concept, called the River Continuum Concept (RCC), linked the hydrologic and geomorphic processes controlling networks of streams and rivers to processes controlling basic ecosystem parameters such as the ratio of primary production (photosynthetic production) to respiration. A schematic diagram presenting the RCC is presented in Figure 3. This concept has provided a unifying theme for understanding streams and rivers, and new hypotheses have enriched this basic concept. The applicability of the RCC has now been evaluated at a global scale.

The watershed approach has been implemented in many areas of the United States and around the world. In addition to organizations conducting research, the Western Water Policy Review Advisory Commission estimates that there are currently 550 watershed stakeholders' groups in the western United States alone. The RCC was used in the design of the National Water Quality Assessment (NAWQA) program of the U.S. Geological Survey, and the scale of the river basins studied in NAWQA matches that of the RCC. Currently, "ecology" and "watershed" are household words, and the ecosystem concept is taught to schoolchildren, indicative of the impact and relevance of the discipline. If, in the future, we continue to apply other areas of limnology to freshwater issues, the term *limnology* may also become widely recognized.

For a given inland aquatic ecosystem, our understanding of hydrologic and ecological processes varies greatly. Ground water provides a striking example. The science of ground water hydrology is well established, with a broad knowledge base, including detailed physical models that allow quantitative study of ground water flow systems (Freeze and Cherry, 1979). On the other hand, the science of ground water ecology has made great advances in the past 20 years but is still in a "fledgling" state. Much remains to be learned in order to apply ground water ecology to the resolution of ground water contamination problems. The

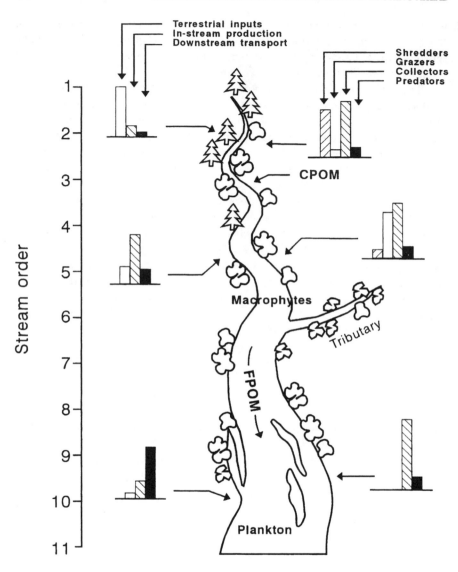

FIGURE 3   Diagram of the river continuum concept depicting a river channel and riparian vegetation as the river grows from a headwater stream to an eleventh-order stream, although the tributaries are not shown.   Source: Johnson et al. (1995) adapted from Vannote et al. (1980).   Reprinted, with permission, from Canadian Journal of Fisheries and Aquatic Science. © Note: Coarse particulate matter (CPOM); Fine particulate matter (FPOM).

current interest in introducing designer microbes to carry out in situ degradation of organic contaminants in ground water illustrates the pressing need for understanding ground water ecosystems.

In the introductory chapter to their book *River and Stream Ecosystems*, Cummins et al. (1995) point out that:

> A hallmark of flowing-water studies during the 1980s and 1990s has been their interdisciplinary nature. The interactions involve biological stream ecologists, hydrologists, geomorphologists, microbiologists, and terrestrial plant ecologists, all interacting to develop generalizations concerning riverine ecosystems.

As a result of the integrative nature of limnology, it may be true that aquatic ecologists have a greater familiarity with hydrology than vice versa. Another reason for this disparity may be that the time scales over which hydrology controls aquatic ecosystems are so long that these controls are not observed by hydrologists during short-term field studies. However, now that dams have been built and hydrologists are involved in studies of geomorphic and ecological changes associated with the modified flow regimes and lowered water tables, this situation is likely to change. The recent experimental flood in the Grand Canyon is an exciting example of collaboration between hydrologists and ecologists that has led to better understanding of these long-term interactions (Collier et al., 1997).

Definition of an Aquatic Ecosystem

A new term has emerged to describe a more holistic approach toward management of natural resources—*ecosystem management*. This term implies a more comprehensive approach than those of the past, which appear piecemeal in retrospect. One of the fundamental concepts of ecology is the ecosystem, broadly defined as an assemblage of species and the dynamic physical environment that they inhabit (Golley, 1993). The ecosystem concept is most useful because the boundaries of the ecosystem (which encompass the control volume) are flexible and can be drawn to address the specific question being asked (Odum, 1953). Thus, an aquatic ecosystem can be defined as a drop of water on a leaf, a small puddle in a forest, a shallow aquifer, or the entire drainage basin of the Mississippi River. There is no right or wrong dimension to an aquatic ecosystem; the specification of the boundaries is only judged relative to the scientific question or the resource issue at hand. If the question is the loss or gain of solutes as rainwater moves through a forest canopy, then defining the ecosystem as the water pooled on a leaf may be useful (Sollins et al., 1980). If the question at hand is the management of flooding and water quality issues in the Mississippi River, consideration of the entire basin may be necessary (Bayley, 1995).

The boundaries serve to distinguish between the flux of organisms and constituents into and out of the ecosystem and interactions and feedback processes occurring within the ecosystem. Materials produced within the ecosystem are "autochthonous," and materials entering from outside the ecosystem are "allochthonous." In deciding how to set the boundaries, a rule of thumb is that once an organism or constituent has exited across a boundary it no longer influences processes occurring within the ecosystem. This is a "gone is gone" criterion, corresponding to organisms or constituents having dripped off the leaf or having been discharged into the Gulf of Mexico. The discrete nature of a lake or an upland stream is conducive to applying ecosystem boundaries, and the directionality of flow is another aspect of aquatic ecosystems that can make them straightforward to define compared to terrestrial ecosystems.

However, the flow-through nature of aquatic ecosystems can present conceptual challenges. To understand these challenges, the temporal dimension of an aquatic ecosystem must be considered as well as the physical dimensions. The hydrologic time scale is determined by the movement of water through the control volume, represented by the residence time. There are also time scales associated with other changes in the physical environment, such as the solar cycle (which drives photosynthesis, evapotranspiration, and other biogeochemical processes). For chemical and biological processes, time scales may vary by orders of magnitude. The time scales of chemical processes depend upon the properties of the major solutes, mineral, and gas phases, as well as the kinetics of important chemical reactions. Within the diverse biota of an aquatic ecosystem, biological time scales range from doubling times of less than days for bacteria and algae to the much longer, more structured life histories of the aquatic plants, invertebrates, and fish. In general, the hydrologic residence time of water, which controls the transport of solutes, organisms, and suspended particulates, may be short relative to the time scales of dominant ecosystem interactions between organisms or between organisms and their environment. As a result, in many aquatic ecosystems the flux of water, material, and organisms moving through the control volume in a unit of time characteristic of important processes may be large relative to the quantity contained within the control volume.

Another challenge is that in inland aquatic ecosystems the important interactions often involve stationary organisms and moving chemical constituents. In streams, for example, algae growing on rocks (periphyton) take up dissolved nutrients from the overlying water and some benthic invertebrates living on the streambed feed on suspended organic material in the overlying water. Similarly, in ground waters biogeochemcial conditions are controlled by bacteria attached to aquifer materials as biofilms. Because of the interactions between moving and stationary components of the ecosystem, the problem of high water flux through the control volume of an aquatic ecosystem cannot be solved by using a frame of reference that moves with the flow. The interactions between flow regime, stream habitat, and habitat requirements for important aquatic species are central to implementation of ecosystem management approaches for streams and rivers (Richter, 1995).

Examples of Hydrology Controlling Aquatic Ecosystems

There have been exciting conceptual advances in aquatic ecology in the past two decades. Many of these advances have been made by quantitatively addressing the hydrologic interactions occurring at ecosystem boundaries (ecotones). These advances would not have occurred without the active participation of hydrologists. In the following sections, examples from current limnological research are presented that illustrate the role of hydrology in defining aquatic ecosystems. The examples begin at the largest scale, that of large river systems, and continue to finer scales, such as the transitional zones in ground waters.

The Comprehensive River Continuum Concept

It is beyond the scope of this paper to discuss in detail the important ideas that make up the more comprehensive River Continuum Concept. In general, these additions have refined our understanding of the hydrologic linkages occurring (1) in the water column and streambed sediments (such as nutrient spiraling and the hydraulic food chain model) (Power et al., 1995), (2) within the floodplain (flood pulse hypothesis), and (3) within the hyporheic zone (the underlying substrate near the stream where water is also flowing in the downstream direction). Only the nutrient spiraling concept and the flood pulse concept will be described to illustrate critical hydrologic interactions.

Nutrient spiraling is a concept that integrates the flow-through aspect of streams with the important biogeochemical role of the substrate and organisms attached to the streambed (Newbold et al., 1981). A given molecule present as a solute in flowing water is converted to a particulate form as it is taken up by streambed biota. Molecules thus taken up are then released from the same biota via excretion or death and returned as solutes to the flowing water, resulting in a nutrient cycle or spiral. For specific molecules of interest, the rate of biological processes in the streambed is compared to the hydrologic transport processes to determine the overall rate of nutrient cycling. The nutrient spiraling concept allows one to assess how stream ecosystems will respond to different flow and substrate conditions (see Figure 4).

In the flood pulse hypothesis the seasonal advance and retreat of water on the floodplain is seen as an important environmental feature to which the biota have adapted (see Figure 5) This hypothesis emphasizes the connection of the stream ecosystem to its floodplain, and implicates flooding as a major control on aquatic biota such as fish. Aquatic ecosystems in the United States have experienced considerable change as a result of flow regulation and other hydrological alterations in rivers. Controls are so pervasive in many rivers that natural, wild rivers are few and far between. In the "Freshwater Imperative" research agenda (Naiman et al., 1995), the topic of understanding "modified hydrologic regimes" was assigned a high priority. The recent successful experimental flood in the

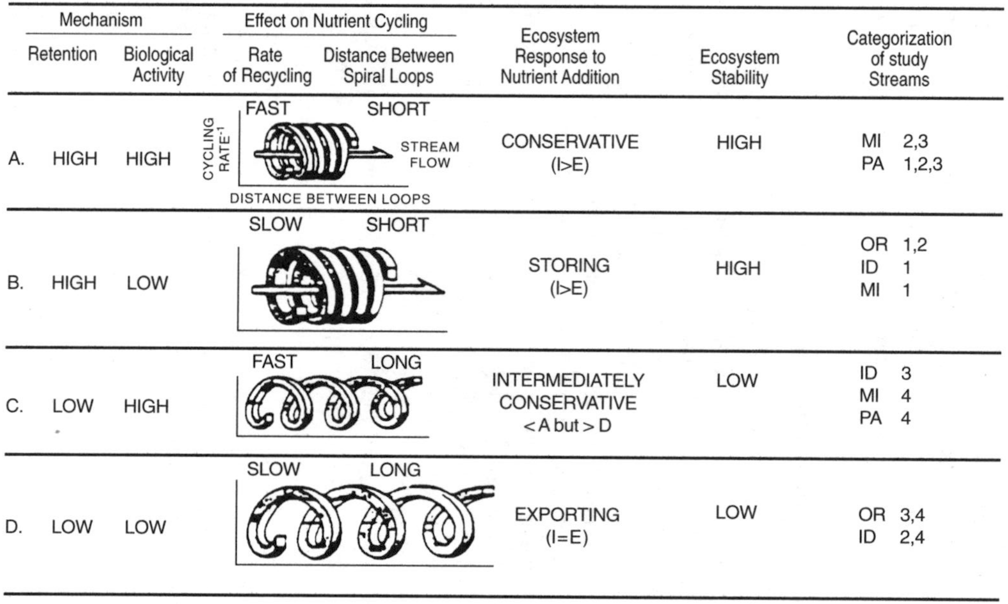

FIGURE 4 Schematic showing the concept of nutrient spiraling in streams. Source: Reprinted, with permission, from Minshall et al. (1983). © by the Ecological Society of America. Note: Oregon (OR), Idaho (ID), Michigan (MI), and Pennsylvania (PA).

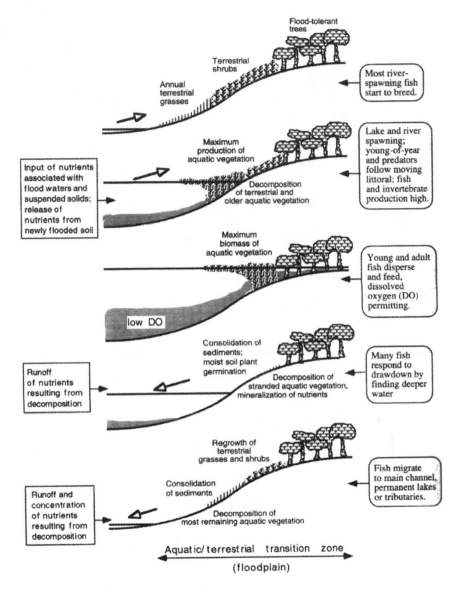

FIGURE 5   Schematic diagram illustrating the floodplain pulse concept, which emphasizes the interaction between river and floodplain in relation to fish behavior. Source: Reprinted, with permission, from Bayley (1995). © 1995 by Bioscience.

Grand Canyon has provided scientific understanding that will support new approaches for regulating flow to achieve environmental and power generation

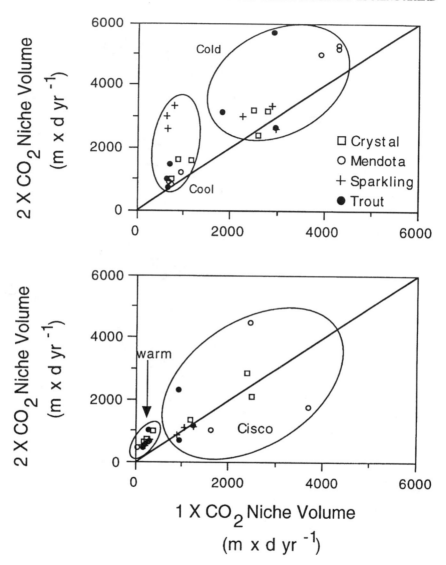

FIGURE 6   Potential effect of climatic changes from a doubling of atmospheric carbon dioxide of the niche volume for fish in selected temperate lakes.   Source: Reprinted, with permission, DeStasio et al. (1996). © 1996 by Limnology and Oceanography.   Note: (m) month and (d) day.

goals. Such approaches can be used across the country when modifying flow regimes on regulated rivers to gain healthier stream communities.

Lake Ecosystems and Climate Change

Temperate lakes respond to climatic forcing. Thus, research on lake ecosystems should be included in research on human-accelerated environmental change. Current studies have considered both how lakes are changing in response to the changing climatic conditions of the last century and how lake ecosystems may change if there is a directional shift in the climate driven by the buildup of carbon dioxide or its radiative equivalent in the atmosphere. Both of these considerations have generally led to new insights into the functioning of lake ecosystems, especially in the context of wintertime processes for which fewer field measurements of dynamics have been carried out.

Circulation and stratification of physical properties in lakes are controlled by temperature, wind speed, and the clarity of the water. These characteristics in turn can influence the distribution of organisms and ecological interactions. Fish populations provide one example of how temperature regimes in the lake can define the habitat for species of fish with different thermal tolerances. DeStasio et al. (1996) used climate predictions from a global circulation model to provide climate parameters mimicking a doubled carbon dioxide atmosphere for a set of four lakes in Wisconsin. They predicted the new summer "niche volumes" for fish species tolerating cold, cool, and warm water temperatures. Because of increased onset of stratification, increased epilimnetic summer temperatures, and a longer duration of stratification, suitable thermal habitats were more abundant for all fish types and all climate change scenarios considered, as shown in Figure 6. The modeling exercise also showed that surface water temperatures may exceed upper lethal limits for some warm and cool water fish. The timing of ice cover is another climate-driven parameter that strongly influences the annual ecological cycle in temperate lakes. Primarily from newspaper accounts, long-term records of ice-on and ice-out dates are available for many lakes in the central United States. Analysis of these records shows a progressive trend of decreasing ice-out dates; an example of a typical record is shown in Figure 7. The regional trend of the duration has also been studied using remote sensing data, one of the concrete ways to quantify the changes in freshwater ecosystems that are driven by changing climatic conditions.

Beyond the observed trends in the duration of ice cover, a fundamental physical understanding of the processes linking climate and ice cover is critical to predicting the changes in ice cover that might occur in response to the buildup of carbon dioxide in the atmosphere. In turn, these changes could be used as input to ecosystem response models. Physical limnologists have been making progress in this area of research, and the schematic diagram in Figure 8 illustrates the

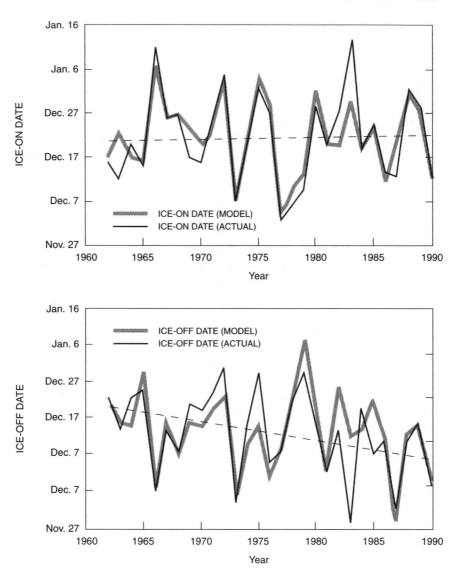

FIGURE 7 Time series of simulated and observed ice-on and ice-off dates for Lake Mendota. Source: Reprinted, with permission, from Vavrus et al. (1996). © by Limnology and Oceanography.

different energy flux terms representing the climate/ice cover linkage in the LIMNOS numerical lake ice model (Vavrus et al., 1996). By using sensitivity analysis, it was shown that the ice-off date was more sensitive to atmospheric

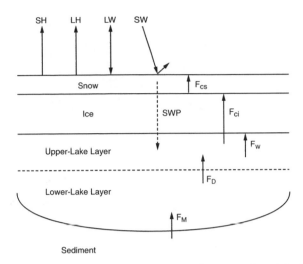

FIGURE 8   Schematic diagram showing energy flux terms in a lake ice model. Reprinted, with permission, from Vavrus et al. (1996). © 1996 by Limnology and Oceanography. Note: SH–sensible heat flux; LH–latent heat flux; LW–net longwave radiation flux; SW–net solar radiation flux; SWP–penetrative solar radiation flux; F_{ci}–conductive heat flux through ice; F_{cs}–conductive heat flux through snow; F_w–Basal heat flux from lake to ice bottom; F_D–eddy-diffusion heat flux; F_M–heat flux from sediment.

temperature than the ice-on date and that increased snow cover can cause a delay in the ice-out date. Thus, temperatures and precipitation measurements during both winter and spring are important parameters for climate simulations.

Lake sediments are integrative records of processes occurring in a given year. Sediments contain sensitive records of conditions in the surrounding watershed and water column. Various biological, physical, and chemical indicators can be interpreted together to construct the past climate and hydrologic conditions in a watershed. These efforts are aided by the tools of paleolimnology. In particular, knowledge of the habitat range of diatom and chrysophyte fossils has been useful. This is an example in which greater biological knowledge (e.g., improvements in the taxonomy of diatoms) can be instrumental to improvements in paleohydrology and applied areas of hydrology such as hazard assessment.

Ground Water Ecosystems

Ground water is the largest reservoir of liquid freshwater in the world. It is an important water resource that is affected by a range of human activities (Freeze and Cherry, 1979). Although the influence of microorganisms on the geochemistry of ground water has been recognized for some time, it has only been in the

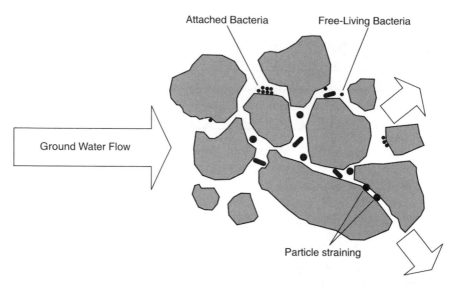

FIGURE 9   Diagram showing the relationship between bacteria and the porous media in groundwater ecosystems. Source: Reprinted, with permission, from Smith and Garabedian (1998). © from Hall and Chapman.

past several decades that ground water has been studied as an aquatic ecosystem (Ghiorse, 1997). Ground water ecosystems differ from surface water ecosystems in that the surface area of the solid phases is very high relative to the volume of the water. Microbial populations responsible for influencing the ground water chemistry reside on these surfaces, except for a small portion of free-living bacteria and viruses that are transported by ground water flow. The distribution of microorganisms in sediments is represented schematically in Figure 9. Ground water ecosystems are inherently different from surface water ecosystems because photosynthesis is not possible. Chemotrophic processes are certainly important in some ground waters, and are generally limited by the availability of substrates such as carbon sources and terminal electron acceptors.

As emphasized in the Introduction, measuring the rates of important processes is critical when studying an aquatic ecosystem. This is particularly challenging in ground water for two reasons. One is the inaccessible nature of the ground water ecosystem itself. Often in ground water studies the greatest expense in terms of field resources is the drilling of the wells. This is a common experience among ecologists, chemists, and hydrologists who are studying ground water systems. In fact, the cost saving in using an existing well may be a factor that promotes interdisciplinary collaboration. Even when core material has been retrieved under the most careful conditions, the disturbances in the fine-scale structure of the material (see Figure 9) are bound to be large. These disturbances are large enough to influence the rates of microbial processes during incubation experiments, in which the active microbial populations within the core material are exposed to different substrates or contaminants.

Rate measurements are also problematic in ground water ecosystems because ground waters are typically oligotrophic. The concentrations of substrates are very low, causing the microorganisms to grow at very slow rates. Even though the water flow rates are much slower than in surface waters, these flow rates can still be substantial compared to the growth rate of a bacterial cell in the interstices of a sand grain (Smith and Garabedian, 1998). In addition to bacteria and viruses, protozoan grazers occur in ground water ecosystems, and the rates of grazing relative to the growth rates of the bacterial populations are very hard to measure or estimate.

With the advance of ground water ecology, the initial view of the subsurface habitat as uniform was replaced by a three-dimensional view of the subsurface, which contains biogeochemical gradients of contaminants. It was recognized that these gradients were controlled by microbial oxidation/reduction processes, in which the sequence of electron acceptors used by degrading microorganisms corresponds to a spatial sequence in biogeochemical conditions in the subsurface (Smith, 1996).

Through collaborations with ground water hydrologists, the methods developed to study ground water flow have been extended to determine the in situ rates of microbial processes. The basic approach is to quantify the hydrology using conservative tracers and solute transport model, and then evaluate the rate of a microbially mediated reaction by comparing the conservative tracer simulations with simulations that include microbial processes. Figure 10 shows the arrival of bromide and methane in an experiment conducted to study microbial oxidation of methane (Smith and Garabedian, 1998). The experiment took place at the Otis Air Force base research site, which is a sand and gravel aquifer contaminated with dilute, treated sewage. In this experiment, methane was coinjected with bromide into the aquifer, and the attenuation of the methane concentration relative to the bromide allowed for quantification of the oxidation rate of methane in this aquifer. Although such field experiments are quite resource intensive, they provide for the coupling of the hydrologic and microbial processes, yielding more definitive information about the process rates in ground water ecosystems. These field-scale experiments are a critical tool for studying ground water contamination and designing in situ ground water remediation.

Toward "Knowledge-Based" Consensus: Future Science and Information Needs

In the introductory discussion the link between water resources and social progress was emphasized. This link has been operational in the United States during this century and has also been maintained by advances in aquatic science

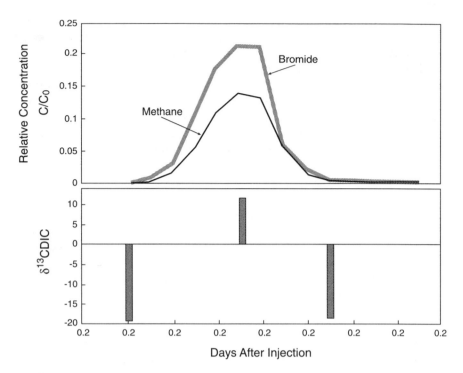

FIGURE 10   Comparison of the arrival of a conservative tracer, bromide, and methane, which was undergoing microbial oxidation.   Source: Reprinted, with permission, from Smith and Garabedian (1998).   © from Hall and Chapman.

and technology. In many regions of this country, city planners projected expanding populations for their cities and developed large systems of reservoirs to provide water supply for the future. Their foresight has benefited the current metropolitan residents, who have derived both clean water and recreational use from man-made reservoirs. Also in this century, there have been significant improvements in drinking water quality through the chlorination of drinking water, resulting in an overall increase in public health. Regulations coming into effect in the next few years are more sophisticated, addressing the production of harmful chlorinated organic compounds during chlorination of drinking water.

While advances were made in some areas of water resources in the mid-century, the quality of many surface waters deteriorated because of municipal and industrial pollution. Implementation of federal legislation has resulted in significant improvements in water quality in many rivers. For numerous Midwestern cities the revitalization of the downtown areas has been centered on developing riverside parks where residents and weekend visitors from the suburbs can stroll, jog, or rollerblade alongside a river that was once seriously contaminated. Although restoration of wetlands and remediation of ground water have been more limited, similar benefits to the public good are anticipated.

What will be required to continue this progress as pressures on water resources build? Many competing interests will be involved in resolving these pressures, and the involvement of stakeholders at the local and regional levels will result in decisions that are specific to a particular aquatic ecosystem. Generic water quality standards will be replaced by site-specific or regional aquatic life criteria. Recent first steps in regional assessments include the report on aquatic ecosystems to the Western Water Policy Review Advisory Commission (Minckley, 1997) and the regional assessment of the impacts of climate change in North America (Cushing, 1997). Federal legislation and regulations alone will not be adequate. Greater scientific knowledge, a more broadly trained cadre of aquatic scientists and engineers, and sufficiently detailed and accurate monitoring networks and databases for the nation's freshwater resources are necessary.

Educational and Professional Issues

Recent assessments of the status of education in limnology have identified major improvements needed for the future (Wetzel, 1995; National Research Council, 1996). Over the past quarter century, limnology has grown into a truly integrative science, fulfilling the vision of the founders of limnology. With this growth, limnology has become dispersed among the academic departments in a manner that varies greatly among institutions. This fragmentation of limnology courses within a university decreases the probability that a student will have easy access to course work that would form a solid foundation for a career in limnology. Further, the lack of emphasis in limnology by one department may translate into inadequate support for laboratory and field training for undergraduate students studying limnology. The problem of availability of courses is especially acute at the undergraduate level. At many universities, undergraduates were not able to take limnology courses because the enrollment demand could not be accommodated by resources assigned for instruction in limnology.

The lack of breadth in graduate training in water-resource-related fields is also a concern. For example, it is not uncommon in civil and environmental engineering departments for students to be required to master hydrology and water chemistry but to have no requirement to learn aquatic ecology. Students may not encounter ecosystem management approaches and aquatic life criteria as targets in remediation until after graduation. The Committee on Opportunities in the Hydrologic Sciences made recommendations to improve education in limnology by establishing regional centers of limnology through the creation of strong aquatic science departments offering comprehensive undergraduate majors in limnology and by establishing strong interdepartmental programs with an opportunity to specialize in limnology. Other recommendations centered on strengthening limnology as a profession through certification programs for limnologists offered by professional societies and establishing a limnologist job classification within the federal government's hiring system. This would help to define for water resource managers the types of training that would be useful in implementing an ecosystem management approach for a water resource.

Aquatic Ecosystem Monitoring and Characterization

As has been emphasized in this paper, the important way in which hydrology defines aquatic ecosystems is in defining the flow of water, solutes, organisms, and detritus through the ecosystem over a range of temporal and spatial scales. Hydrochemical data are a key "raw material" in the coupling of aquatic ecology and hydrology. Just as long-term hydrologic records have been important in designing reservoir systems for flood control and water supply, long-term hydrochemical records such as for the Hubbard Brook watershed are important in understanding water quality trends. Thus, in order to "take an ecosystem approach," some relevant hydrologic and chemical information is needed about a given aquatic ecosystem or a similar nearby ecosystem. The specific data needs are dependent upon the specific question or issue. The willingness of the stakeholders to participate in a management plan may be influenced by the quality and the credibility of the database upon which the ecosystem analysis is based.

For these reasons now is not the time to be consolidating and merging monitoring networks and degrading the geographical resolution of the current networks. The seriousness of this issue is reflected in the decreasing trend in the number of stream-gaging stations operated by the U.S. Geological Survey; in 1990, 7,400 stations were operated, and in 1997 the number of stations decreased to 6,800 (Carlowicz ,1997). It is time to protect and preserve existing monitoring networks that provide basic data on flow and chemistry with a degree of reliability that has established the network as a source of credible data. For a "knowledge-based" consensus to take hold among different competing interests involved in a water resource issue, some hydrologic records are required to form a basis for interpretation of other information.

SUMMARY

Aquatic ecosystems are defined by hydrologic processes. The great advances that have been made in limnology in the past several decades have been fueled by interactions with hydrologists. Research programs in limnology should be greatly expanded such that the scientific benefits of these interactions can be realized in a timely manner. Through expanded research, we will develop a more holistic approach towards addressing the freshwater issues of the future. Thus, the interaction of limnologists and hydrologists should be encouraged, and continued efforts should be made to influence members of the engineering community who discourage the participation of limnologists (such as Lyon, 1997).

To foster this ongoing collaboration, we must strengthen undergraduate and graduate education in limnology. In addition to changes within the universities, establishment of a "limnologist" professional track in the federal and state governments would help to advance limnology as a profession. It is also important to continue the long-term monitoring programs that provide the hydrologic and chemical data needed for ecosystem management. We should not let the budget ax fall now when these data can be used effectively to achieve "knowledge-based consensus" in the management of specific freshwater resources.

ACKNOWLEDGMENTS

I acknowledge helpful discussions with D. Niyogi, A. Brown, R. Smith, J. Cole, J. Baron, and G. Hornberger, which provided ideas and insight for the manuscript.

REFERENCES

Bayley, P. B. 1995. Understanding large river-flood plain ecosystems. Bioscience 45:153-158.

Carlowicz, M. 1997. USGS retrenches, regroups after weathering budget hazards. EOS 78(16):165-168.

Collier, M. P., R. H. Webb, and E. D. Andrews. 1997. Experimental flooding in Grand Canyon. Sci. Am. 276(1):82-89.

Cummins, K. E., C. E. Cushing, and G. W. Minshall. 1995. Introduction: An overview of stream ecosystems. In River and Stream Ecosystems. Ecosystems of the World Volume 22. Amsterdam, New York: Elsevier.

Cunliffe, B. 1993. The Roman Baths: A View over 2000 Years. Bath, England: Bath Archeological Trust.

Cushing, C. E. 1997. Fresh Ecosystems and Climate Change in North America: A Regional Assessment. New York: Wiley.

DeStasio, B. T., Jr., D. K. Hill, J. M. Kleinhans, N. P. Nibbelink, and J. J. Magnuson. 1996. Potential effects of global climate change on small north-temperate lakes: Physics, fish, and plankton. Limnol. Oceanogr. 41:1136-1149.

Freeze, R. A., and J. A. Cherry. 1979. Groundwater. Englewood Cliffs, N.J.: Prentice-Hall.

Ghiorse, W. C. 1997. Subterranean life. Science 275:789-790.

Golley, F. B. 1993. A History of the Ecosystem Concept in Ecology. Binghampton, N.Y.: Vail-Ballou Press.

Johnson, B. L., W. B. Richardson, and T. J. Naimo. 1995. Past, present, and future concepts in large river ecology. Bioscience 45(3):134-141.

Likens, G. E., F. H. Bormann, R. S. Pierce, J. S. Eaton, and N. M. Johnson. 1977. Biogeochemistry of a Forested Ecosystem. New York: Springer-Verlag.

Lindeman, R. L. 1941. Seasonal food-cycle dynamics in a senescent lake. Am. Midl. Nat. 26:636-669.

Lyon, W. A. 1997. A research agenda or an ideological treatise? Review of the Freshwater Imperative. Rivers 5(4):304-307.

Mayr, E. 1982. The Growth of Biological Thought. Cambridge, Mass.: Harvard University Press.

Minckley, W. L. (ed). 1997. Aquatic Ecosystems Symposium. A report to the Western Water Policy Review Advisory Commission. Arizona State University, Tempe.

Minshall, G. W., R. C. Peterson, K. W. Cumminis, T. L. Bott, J. R. Sedell, C. E. Cushing, and R. L. Vannote. 1983. Interbiome comparison of stream ecosystems dynamics. Ecol. Monographs 53:1-25.

Naiman, R. J., J. J. Magnuson, D. M. McKnight, and J. A. Stanford. 1995. The Freshwater Imperative: A Research Agenda. Washington, D.C.: Island Press.

National Research Council. 1996. Freshwater Ecosystems: Revitalizing Educational Programs in Limnology. Washington, D.C.: National Academy Press.

Newbold, J. D., J. W. Elwood, R. V. O'Neill, and W. VanWinkle. 1981. Measuring nutrient spiraling in streams. Can. J. Fish. Aquat. Sci. 38:860-863.

Odum, E. P. 1953. Fundamentals of Ecology. Philadelphia: W. B. Saunders.

Pool, R. 1991. Science literacy: The enemy is us. Science 251:266-267.

Postel, S. 1997. Dividing the waters. Technol. Rev. 100(3):54-62.

Power, M. E., A. Sun, G. Parker, W. E. Dietrich, and J. T. Wooton. 1995. Hydraulic food-chain models. Bioscience 45:159-167.

Richter, B. D. 1995. Integrating science in applied aquatic ecosystem management. Bull. NABS 12(1):82.

Smith, R. L. 1996. Determining the terminal electron-accepting reaction in the saturated subsurface. In Manual of Environmental Microbiology. C. J. Hurst, G. R. Knudsen, M. J. McInerney, L. D. Stetzenbach, and M. V. Walter, eds. Washington, D.C.: ASM Press.

Smith, R. L., and S. P. Garabedian. 1998. Using transport model interpretations of tracer tests to study microbial processes in ground water. Pp. 94-123 in Mathematical Modeling in Microbial Ecology, A. L. Loch, J. A. Robins and G. A. Milkier, eds. New York: Chapman and Hall.

Sollins, D., C. C. Grier, F. M. McCorison, K. Cromack, Jr., R. Fogel, and R. L. Fredriksen. 1980. The internal element cycles of an old growth Douglas-fir ecosystem in western Oregon. Ecol. Monogr. 50:261-285.

Vannote, R. L., G. W. Minshall, K. W. Cummins, J. R. Sedell, and C. E. Cushing. 1980. The river continuum concept. Can. J. Fish. Aquat. Sci. 37:130-137.

Vavrus, S. J., R. H. Wynne, and J. A. Foley. 1996. Measuring the sensitivity of southern Wisconsin lake ice to climate variations and lake depth using a numerical model. Limnol. Oceanogr. 41:822-831.

Wetzel, R. G. 1995. Training of aquatic ecosystem scientists: Continue to languish or accept our responsibilities? Water Resour. Updates 98:15-20.

3

Hydrologic Measurements and Observations: An Assessment of Needs

Eric F. Wood
Department of Civil Engineering
Princeton University

INTRODUCTION

No one believes a model save its developer.
Everyone believes a data set except its collector.
—Anonymous

Hydrology is a science built on observations and measurements. Hydrologic theories either have emerged from insights gained through analyzing data or have been confirmed through data that support the theory. The report *Opportunities in the Hydrologic Sciences* (NRC, 1991) recognizes the importance of data collection, distribution, and analysis by devoting a chapter to the issues concerned with this critical topic. The chapter presents compelling arguments for (1) the *need* to continue to utilize observational networks and experimental measurements, (2) an assessment of the *status* of hydrologic data collection, and (3) exploiting *opportunities* to improve hydrologic data. It seems redundant to repeat the arguments here. Nonetheless, it is possible to provide an assessment of whether concerns raised in the 1991 report have been heeded and whether opportunities have been seized that could provide new, innovative measurements for hydrological theory.

The introduction to that chapter states that:

> . . . today there is a schism between data collectors and analysts. The pioneers of modern hydrology were active observers and measurers, yet now designing and executing data collection programs, as distinct from experiments carried out in a field setting with a specific research question in mind, are too often

viewed as mundane or routine. It is therefore difficult for agencies and individuals to be doggedly persistent about the continuity of high-quality hydrologic data sets. . . . The scientific community tends to allow data collection programs to erode.

How is it that observations and measurements have become the stepchild of the science? Even in the chapter titled "Critical and Emerging Areas," there is a preference toward modeling and theoretical developments over data analysis and testing of current theory. This review will look at three related questions in the area of measurements and observations: (1) Has *Opportunities in the Hydrologic Sciences* had an impact in the area of data collection? (2) Are there areas in which data collection has fallen down? (3) Are there areas where data needs are not being met? The review will be from the author's perspective and for the most part will draw examples from terrestrial hydrology.

DATA COLLECTION AND OPERATIONAL HYDROLOGY

In the United States, data collection in support of operational hydrology and water resources goes back more than 100 years. The needs can be described by considering the aggregated water balance equation over a time interval Δt:

$$\Delta S = P - E - Q \tag{1}$$

where S is the change in soil moisture, P is precipitation, E is evapotranspiration, and Q is runoff. In operational hydrology a major concern is flood forecasting. For floods large enough to put the public at risk, the soil tends to be saturated, so ΔS is rather minor and E is negligible, which then leaves the observational requirement of the accurate estimate of precipitation and the timing of the discharge Q. This observational need has had two important manifestations: (1) the development of operational networks operated by the U.S. Geological Survey (USGS) (for streamflow) and the National Oceanic and Atmospheric Administration (NOAA) (for precipitation) and (2) disinterest in establishing observational networks for soil moisture and evaporation by these agencies.

Flooding in the United States remains a serious problem that results in significant loss of life and much damage. Figure 1 illustrates that these events are frequent and widespread across the nation. Irrespective of this, there appears to be limited effort by the USGS and the National Weather Service (NWS) to demonstrate, on an economic and social basis, the benefits of an extensive and accurate observational system in support of flood forecasting and warning and to show the relationships among observations, forecasts and warnings, and reduced flood damages and less loss of life.

The National Research Concil recently assessed the hydrological operations and services provided by the NWS (NRC, 1996), and Mason and Yorke (1997)

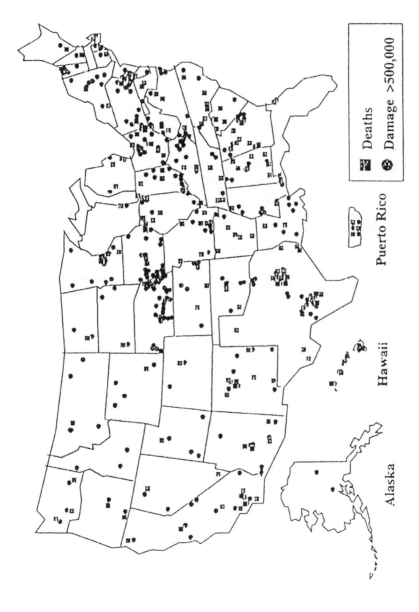

FIGURE 1 Notable floods and flash floods in the United States, 1987-1991. Source: National Weather Service (1992).

have summarized streamflow gaging by the USGS. Without belaboring the point, these assessments support what hydrologists have known for some time: observational networks have degraded over the past 10 years, and the accuracy of observations can be improved. Some specific examples to illustrate these points follow.

Stream Gaging

In 1996 the USGS gaged streamflow at approximately 7,000 stations, a decrease of 185 sites from the previous year and 363 from 1990, indicating an acceleration in site closings. On the other hand, about 60 percent of the sites use satellite telemetry to broadcast stages 24 hours a day to such users as the NWS. Such data are vital to the NWS's mission to provide flood forecasts and warnings. Figure 2 illustrates trends in the USGS's gaging program. What are the strengths? Improvements in the timely delivery of flood stage information can have significant economic benefits, estimated in the February 1996 Willamette Valley, Oregon, floods to be $2.7 billion (Mason and Yorke, 1997). What are the weaknesses? In 1996 the USGS streamflow-gaging program cost the nation $82 million and involved a partnership between the USGS and more than 700 federal, state, and local agencies. While the USGS views the partnerships as a strength, they leave the program vulnerable to the whims of funding from agencies that often view paying for data collection as secondary to other missions in times of shrinking resources.

An example is the removal of the gage on the Licking River in McKinneysburg, Kentucky (near Falmouth) in September 1994, which hindered the delivery of accurate flood stage information and contributed to increased loss of life and damages during the March 1997 floods. Neither this gage or its closing was particularly unique. The McKinneysburg gage, used by the NWS for flood forecasting, had been in operation for more than 50 years. It was one of about 15 gages discontinued in 1994 out of a network of 100 USGS gages in Kentucky. As with almost all of the gages in the network of 7,000 USGS streamflow-gaging stations nationwide, this gage was funded as a partnership of multiple agencies and was operated by the USGS. In this case, 25 percent of the costs were funded by the USGS, 25 percent by the Kentucky Division of Water, and 50 percent by the U.S. Army Corps of Engineers.

Network priorities are regularly reevaluated as available funding and priorities change. In 1994 the agencies that supported the McKinneysburg gage determined that other gages were more critical to the missions of the funding partners and that funding would be withdrawn. The USGS and the NWS sought, to no avail, other funding sources to continue operating the gage. Operation of the gage was discontinued in September 1994. As a reference, since 1990 there have been 182 gaging stations closed that were part of the flood-forecasting network of the National Weather Service.

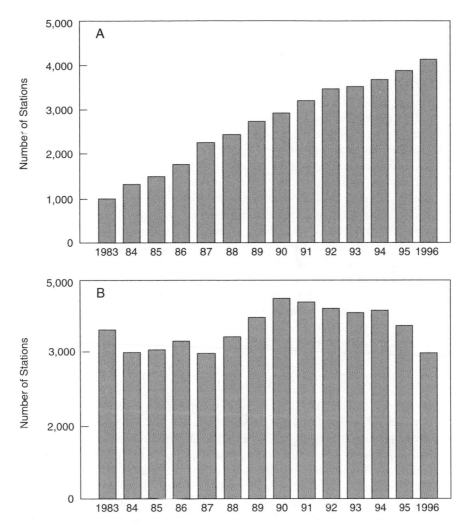

FIGURE 2   Streamflow-gaging stations. A, with satellite data collection platforms; B, total number by year.

Consider another example of how fiscal pressures are affecting data collection. In *Opportunities in the Hydrologic Sciences*, the need to develop accurate long-term hydrologic data bases to improve scientific understanding is highlighted. An example is the development of a network of stream gages and water quality sampling stations in undisturbed basins—index stations—that would be benchmarks against which changes in rivers caused by climate change and human development could be evaluated. Such an effort to set up this network is described as "farsighted" in *Opportunities in the Hydrologic Sciences*. How has

the network fared recently? As Mason and Yorke (1997) state: "The most significant reductions [in gaging sites] occurred in the index-stations networks used by
the USGS to monitor and document long-term changes and trends in water availability and quality."

With these observations, the following recommendation is made: *The
federal government's requirement for streamflow data needs to be established
and a consistent and stable funding mechanism must be put into place.* This
recommendation is made in light of a 1996 USGS survey of the agency's state-
based water resources offices, with input from the NWS river forecast offices,
which identified 1,500 sites where additional real-time streamflow gaging
would improve flood forecasting and flood warnings. These sites included (1)
streamflow gaging stations that were previously discontinued, such as the
McKinneysburg gage; (2) locations where new gages could be installed; (3)
gages where new measurement devices should be deployed to improve gaging
accuracy; and (4) existing gages that need new satellite telemetry to enable
them to provide real-time data. The capital costs for the gages are approximately $40 million, annual operation costs would total $15 million, and incor-
poration of the gages into the NWS flood forecast system would be approximately $15 million. Whether these gages (one or all 1,500) are a good
investment for the nation needs to be established within a program that consid-
ers the broad national needs for stream gage data.

Precipitation

In recognition of the importance of precipitation observations in making
accurate flood forecasts and timely warnings, the NWS is going through an
extensive modernization program. Central to this is the installation of doppler
rain radars (WSR-88D). Tremendous strides have been made in observing precipitation systems at fine temporal and spatial scales through the WSR-88D
radars. But recent analyses have demonstrated that the resulting quantitative
precipitation forecasts are often poor and inconsistent between radars viewing the
same precipitation system (Smith et al., 1996a). For example, Figure 3 shows
WSR-88D-based estimates of three important quantities—conditional mean rainfall, probability of rainfall, and mean hourly rainfall as a function of radar range—
for two radars, Twin Lakes (Okla.) and Tulsa (Okla.). The figures on the left are
for the 1994-1995 period, and those on the right are for 1996. Two critical
features can be seen. First, there is a persistent range effect on the order of a 50
percent difference between the highest and lowest values, while what is expected
are estimates that are constant at all ranges. The algorithmic and radar calibration
improvements installed in 1996 did not eliminate this problem. Second, a significant bias exists between the two radars, which are separated by approximately
100 km. Calibration has removed some of this bias, but the difference in *mean*
hourly rainfall is still over 25 percent.

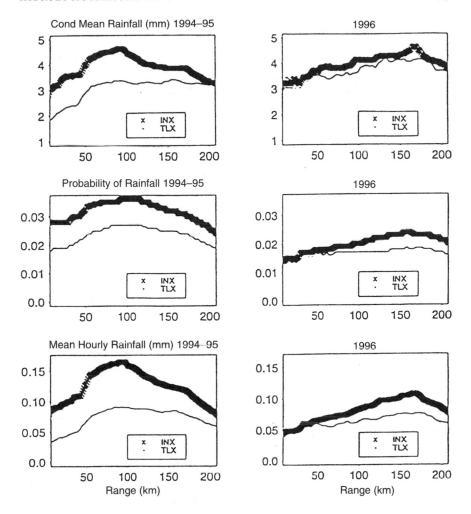

FIGURE 3   WSR-88 radar rainfall estimates as a function of range.   INX refers to Twin Lakes (Okla.) radar and TLX to Tulsa (Okla.) radar.   The figures on the left are based on 1994-1995 data; on the right, 1996 data.   Source: Reprinted, with permission from Bauer-Messmer et al. (1997).   © from American Meteorological Society.

For individual storms, differences between radars can be unacceptably large. Figure 4 shows total storm rainfall from an intense storm over the Rapidan River basin in West Virginia that occurred on July 17-18, 1996. The radar estimates on the left are from the Davenport, Iowa, radar at a range of about 150 km, while the right side of the figure shows estimates from the Chicago, Ill., radar at a range of

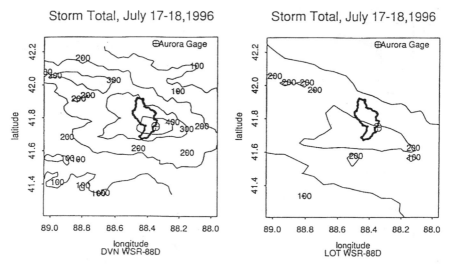

FIGURE 4   Comparison of two WRS-88D total storm rain estimates over the Rapidan River (W.Va.) basin.   Left, Davenport, Iowa.   Right, Chicago Ill.   Source: Smith et al. (1996a).   © 1996 by American Geophysical Union.

about 50 km.   The recording gage at the center of the storm reported 440 mm, which is extremely close to the Davenport radar.   The Chicago radar underreported the rainfall by about 100 percent.   In fact, comparisons between radar-based estimates and gage observations show significant underestimation of heavy rainfall and chronic biases (see Figure 5 from the Tulsa, Okla., radar).

The National Research Council assessment of the hydrological and hydrometeorological services provided by the NWS (NRC, 1996) was quite positive—and rightly so.   But what appeared to be missing in the report were quantitative measures and goals in radar calibration, radar-gage comparisons, and forecast accuracy.   These measures are critical since they are needed to help allocate funding and to evaluate expenditures among competing needs.   This leads to the following recommendation: *There is a need to establish quantitative measures and goals in radar calibration, radar-gage comparisons, and their effect on flood forecast accuracy.*

Operational Modeling Needs

Opportunities in the Hydrological Sciences states that "modeling and data collection are not independent processes. . . . Each drives and directs the other" (NRC, 1991, p. 215).   For the NWS, hydrologic water balance models are indispensable tools for producing site-specific flood forecasts and warnings.   This is

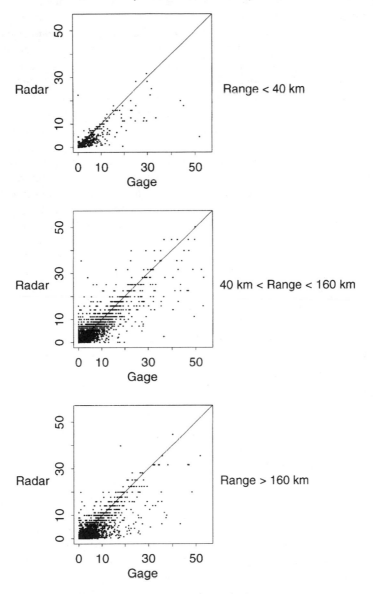

FIGURE 5 Intercomparison of WRS-88D and rain gage rainfall estimates for the Tulsa (Okla.) radar. Source: Smith et al. (1996b). © 1996 by American Geophysical Union.

more evident as localized flash floods account for a greater portion of flood damages and loss of life, owing to expanding populations and increased development. The components of the surface water balance—that is, the partitioning of precipitation into surface runoff and infiltrated soil water, and the partitioning of the soil water into evaporation and transpiration, drainage, and stream discharge— are influenced by surface characteristics such as soil texture, vegetation, and topography. In the past five years the availability of such data has increased tremendously. For example, 1-km soil texture data are widely available, the USGS earth resources observation system (EROS) data center provides weekly advanced very high resolution radiometer (AVHRR) satellite vegetation data, and digital terrain data are available at 90 m nationally and 30 m for about half the country.

The expanding role of operational models for water management and environmental monitoring and prediction has resulted in the need for continuous time (storm and interstorm) modeling of stream discharges and river basin water balances. Considering Eq. (1), this requires that each variable either be specified or represented mathematically based on an understanding of the underlying hydrological processes. This suggests that we need to represent infiltration processes to estimate $\Delta S/\Delta t$; that evapotranspiration must either be measured and used as an input or estimated from the surface energy balance; and that streamflow, which is made up of surface runoff and drainage from the infiltrated water into the drainage network (baseflow), must be accounted for.

For modeling severe flooding, $\Delta S/\Delta t$ and E tend to be negligible, and for the design of water resource projects, climatological estimates of E are sufficient (since design by its very nature represents expected performance). Nonetheless, long-term accurate data sets are needed to determine site-specific constants for operational hydrological models. It is critical that hydrologists continue to articulate this need.

In operational weather forecasting models, such as those used by the National Center for Environmental Prediction (NCEP) or the European Centre for Medium-range Weather Forecasts (ECMWF), the needs are somewhat different. In the past five years there has been a drive to incorporate better representation of terrestrial hydrology, which represents the lower boundary condition to the atmosphere. It has been shown that such improvements lead directly to improved weather forecasts (e.g., Beljaars et al., 1996). The Beljaars et al. (1996) paper is a particularly important one because it reported on the influence of land parameterizations on precipitation forecasts from weather prediction models. The installation and testing of cycle 48 (CY48) occurred during June 1993 and paralleled forecasts produced during July 1993—a period of particularly heavy precipitation and flooding in the central United States. The major improvement to the land surface model included a better representation of the upper soil column moisture and heat flux dynamics—based, to a great extent, on data collected

A   NOAA: Observed July Precipitation

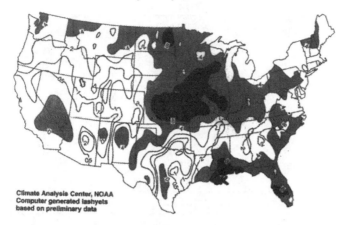

B   Percentage of Normal Precipitation - July 1993

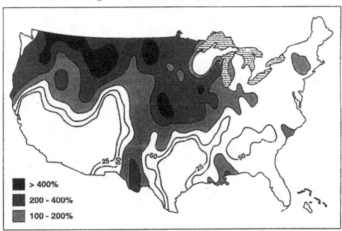

FIGURE 6   Total precipitation over the United States for July 1993 (A) and the percentage of normal precipitation (B) as published by the *Weekly Weather and Crop Bulletin* (August 3, 1993). (A) The contours are at 0.5, 1, 2, 4, and 8 in. with light and heavy shading above 4 and 8 in., respectively; (B) the contours are at 25, 50, 100, 200, and 400 percent, with shading above 100 percent. Source: Beljaars et al. (1996).

under the First international satellite land surface climatology project (ISLSCP) Field Experiment (FIFE) held in central Kansas during the summers of 1987 and 1989 (Sellers et al., 1990, 1993).

Specifically, the operational CY47 system had a land surface scheme that

A CY47

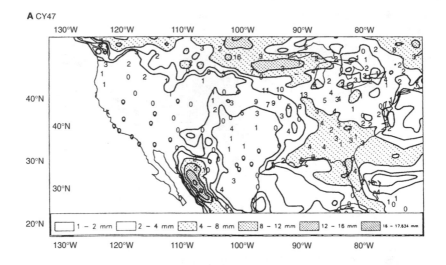

B CY48

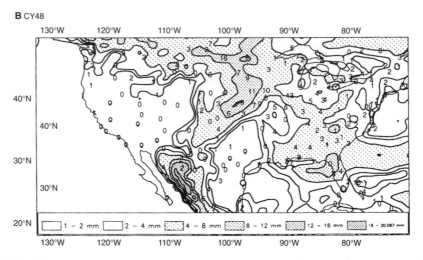

FIGURE 7   Mean forecast precipitation of all 48- to 72-hour forecasts verified between July 9 and 25 with (A) CY47 and with (B) CY48. The contours are at 1, 2, 4, 8, etc., mm/day. The printed numbers are station observations in millimeters per day. Source: Beljaars et al. (1996).

was heavily constrained by a climatologically defined deep soil boundary condition for soil moisture and temperature, whereas in CY48 the model produces a dynamic, time-varying, surface soil moisture through time integration of the atmospheric forcings of precipitation and evaporation. As an example of the impact of the modified land surface representation (summarized by Beljaars et al., 1996), consider Figures 6 and 7. Figure 6 gives the July 1993 total precipitation over the United States as reported in *Weekly Weather and Crop Bulletin*. Figure 7 shows the mean forecast precipitation from ECMWF 48- to 72-hour forecasts from July 9 to 25, 1993. These dramatic improvements of precipitation forecasts—especially in the 48- to 72-hour forecasts—have been the foundation of stated desires by the atmospheric community for extensive soil moisture measurements. In a similar manner, there has been a dramatic improvement in surface air temperature forecasts in high latitudes by incorporating improved winter time albedo measured as part of the ISLSCP Boreal Ecosystem Atmospheric Study (BOREAS) (Viterbo, ECMWF, 1997, personal communication).

These improvements support the stated need in *Opportunities in the Hydrologic Sciences* to develop hydrologic data bases to improve scientific understanding (in this case, improvements in representing underlying hydrological processes and the needed parameters in operational models). Over the past decade the climate experiments carried out under the auspices of the World Climate Research Programme (WCRP)—hydrologic-atmospheric pilot experiment (HAPEX)-MOBILY, HAPEX-Sahel, ISLSCP-FIFE, and ISLSCP-BOREAS—have provided important data that have been used to improve operational weather models. There has been a tremendous improvement in available data to evaluate coupled water and energy balance models (e.g., the improvements in ECMWF's model reported by Betts et al., 1993; Viterbo and Beljaars, 1995; and Betts et al., 1996). But measurements at experimental sites did not start with these programs. The Agricultural Research Service (ARS) of the U.S. Department of Agriculture operated, and still operates, a large number of experimental catchments across the United States. While these experimental catchments had agricultural hydrology as their primary research focus, they collected a wealth of data that potentially could be used to improve the land surface hydrology of operational models.

For example, over the past few years there has been strong interest in soil moisture measurements because of the coupling between soil moisture and evapotranspiration and soil moisture and infiltration—with their subsequent effect on both the energy and the water balances. Current parameterizations of transpiration in operational models (e.g., ECMWF, see Viterbo and Beljaars, 1995, or NCEP's Eta model, see Chen et al., 1996) include functions (Jarvis, 1976) that limit transpiration due to environmental factors, including soil moisture stress. Yet many ARS catchments have long periods of soil moisture measurements in conjunction with precipitation, evaporation, and meteorological data. Over the past few years, the ARS's hydrology laboratory has attempted to organize some of this data into an on-line database (see http://hydrolab.arsusda.gov). There are other important hydrologic data, some collected by the ARS and other agencies, that are sitting in boxes under desks, in field stations, on PCs, and who knows where. Considering the interest in and importance of historical data for improving hydrology models, the following recommendation is made: *An assessment should be carried out of what data are needed to improve the representation of hydrological processes in operational models, the availability of such data, and the effort to incorporate critical data sets into an on-line database.*

MEASUREMENTS AND SCIENTIFIC HYDROLOGY

In the early 1980s there was a recognized need for large-scale climate field experiments to provide the necessary data for achieving progress in (1) coupled land-atmospheric processes for climate modeling, (2) understanding water and energy exchanges at General Circulation Model-grid scales, and (3) utilizing remote sensing data for validating large-scale hydrological models. *Opportunities in the Hydrological Sciences* recognizes these needs and the importance of the synergism between modeling and measurement: "the development of hydrologic theory [and] data collection are interconnected [and] measurement of hydrologic variables is a scientific endeavor itself" (NRC, 1991, p. 215). This section addresses progress made in the past five years to the identified "opportunities to improve hydrologic data" discussed in *Opportunities in the Hydrological Sciences*.

Coordinated Experiments

During the past decade, the field programs developed under the auspices of the WCRP have provided critical data that have advanced scientific hydrology. Recent HAPEX-MOBILY, HAPEX-Sahel, ISLSCP-FIFE, and BOREAS experiments have shown the value of high temporal resolution water and energy flux measurements in evaluating land surface hydrological models (Shao and Henderson-Sellers, 1996; Chen et al., 1996) and making improvements to operational weather prediction models (Betts et al., 1993; Viterbo and Beljaars, 1995; Betts et al., 1996). Carried out in Kansas during the summers of 1987 and 1989, FIFE provided (for the first time) consistent high-quality water and energy flux data suitable for coupled water and energy modeling. This allowed hydrologists to develop and validate coupled water and energy balance models at catchment scales (Famiglietti and Wood, 1994a, b; Liang et al., 1994; Peters-Lidard et al., 1997).

The interactions between experimentalists and modelers, among scientists covering terrestrial hydrology, boundary layer processes, biospheric processes, and remote sensing, were unprecedented. The 764 journal pages in the first FIFE special issue (*Journal of Geophysical Research*, 97(D17), November 30, 1992) are a testament of the level of scientific activity arising from the FIFE data. But the real success of FIFE was National Aeronautics and Space Administration's (NASA) open data policy, in which data are freely available to all from the time of collection. In addition, the effort was made and money was spent to incorporate the data into a database. The availability of these data, first through CDs and currently over the Internet, played a critical role in expanding the scientific impact of FIFE beyond its initial scientists. Following NASA's lead, there has been greater openness and availability in distributing data from other WCRP climate studies.

Two main goals identified from these coordinated studies, stated in *Opportunities in the Hydrologic Sciences*, are (1) understanding the energy and water cycles at scales beyond the field plot and (2) combining the satellite and *in situ* measurements in order to verify large-scale hydrologic models. The observations taken under the WCRP experiments have been of mixed value for large-scale terrestrial hydrology. This is because the major programs had foci that only included terrestrial hydrology peripherally. HAPEX focused on hydrology atmospheric parameterization. ISLSCP was formed in 1993 from interests in remote sensing, climate, and ecological communities to look at the appropriateness of visible/infrared (VIS/IR) satellite data to study land surface *biogeophysical* properties and their role in partitioning incoming radiation into latent and sensible heat. In a similar manner, the formation by ICSU (International Council of Scientific Unions) of the core project on the Biospheric Aspects of the Hydrological Cycle (BAHC) had as its focus ecological and biophysical terrestrial processes. Only through GEWEX, the Global Energy and Water Cycle Experiment, is there a direct goal linked to terrestrial hydrology.

The products of the HAPEX and ISLSCP program goals are data sets of limited usefulness for modeling and validating water and energy balances across a range of scales from field/tower scales to regional and continental scales. The experimental plans emphasized detailed spatial measurements for short periods (which are more suitable for developing satellite retrieval algorithms) over long-term measurements. In addition, there was more emphasis on biophysical and atmospheric measurements than terrestrial hydrologic fluxes. Therefore, carrying out basic analyses such as a water budget during the FIFE experimental period is difficult due to incomplete and uncertain data (Duan et al., 1996). In fact, it is impossible to use data from any climate experiment—including the U.S. Department of Energy's cloud-atmospheric radiation testbed/atmospheric radiation measurement (CART/ARM) experiment—to carry out an observation-only water and energy balance over a catchment or grid of 10,000 km^2 or greater.

As another example, the BOREAS experimental plan made no allowance for accurate measurement of precipitation, and only through separate funding from the National Science Foundation and Environment Canada was a rain radar system installed for the Southern Study Area. Insufficient stream gaging and inadequate measurements of surface infiltration, interflow, percolation, and ground water movements have resulted in unresolved differences between measurements of water and energy fluxes at different BOREAS scales. It appears that the supporting role of hydrology in these experiments resulted in inadequate planning and funding for critical hydrologic measurements. Over the long term, the hydrological data from these experiments should be the most valuable data. As agencies start planning for the new ISLSCP experiment in Amazonia, it is likely that inadequate planning and resources will go into the hydrologic measurements and analyses of hydrologic processes. This deficiency will result in scientists being unable to answer basic questions posed in the scientific plan. This leads to the following recommendation: *There is a need to better define the hydrologic data needed to achieve the stated goals of the WCRP climate experiments, including goals related to atmospheric and biophysical processes, and to strengthen the terrestrial hydrological studies within the WCRP, HAPEX, and ISLSCP climate experiments.*

As an aside, the GEWEX-Mississippi River basin study (GCIP) and GEWEX-Mackenzie have considered the terrestrial hydrologic data needs more carefully. GCIP has performed well in gathering together hydrological data to understand hydrological processes across scales up to the continental watershed scale, and this has resulted in significant improvements in NCEP's operational Eta model.

Nonetheless, the program has relied on operational data whose quality may be insufficient for the goals of GEWEX or of scientific hydrology. Such operational data include precipitation from the WSR-88D rain radar, streamflow from the USGS network, surface meteorology from surface airway stations, atmospheric profiles from NOAA's radiosonde network, and radiation from the geostationary GOES satellite. Each of these have shortcomings in reaching the GEWEX goal "of determining water and energy fluxes by measurements of observable atmospheric and surface properties." As an example, consider Figure 8, which shows accumulated precipitation over a seven-day period in the Red-Arkansas basins. Systematic errors among different WRS-88D radars resulted in identifying different radar footprints. Such errors may have a limited effect on operational forecasts, but they greatly affect the accuracy of long-term water balances. Thus, the following recommendation is proposed: *For GEWEX experiments utilizing operational data, there is a need to establish the accuracy requirements for data related to estimating water and energy balances across a range of scales.*

Remote Sensing

Readily available satellite observations represent one of the most positive aspects of the past decade and have been brought about by a combination of WCRP/ICSU initiatives like ISLSCP, NASA's Earth Observing System (EOS) research program, and advances in computer storage and transmission. The challenge here is not acquiring satellite data but rather processing and transforming the data into needed biogeophysical parameters.

This last step requires broadened ISLSCP-type programs to acquire ground-truth data for the development of satellite retrieval algorithms, and a set of (global) ground validation sites. NASA currently plans to fund a validation program

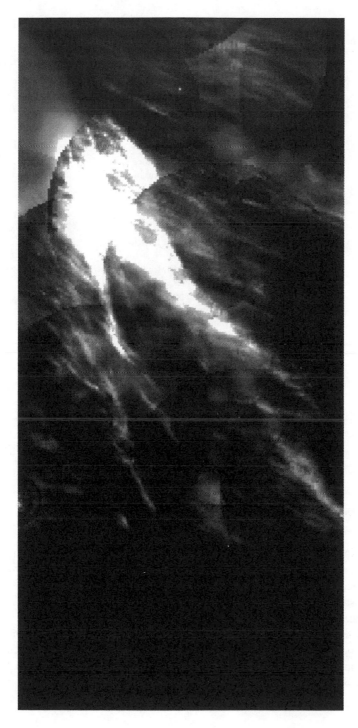

FIGURE 8   WRS-88D accumulated precipitation estimates over the Red-Arkansas river basins region of GCIP for April 8-14, 1994.

TABLE 1   EOS Instruments and Products and How They Will Be Used By Land-Surface Climate Models

Instrument	Products	Uses
ASTER	Land surface temperature, snow cover, cloud characteristics, albedo Visible and near-infrared bands, elevation, albedo	Forcing Parameterization Validation
CERES	Albedo, radiation fluxes, precipitable water, cloud forcing characteristics, surface temperature aerosols, temperature humidity, and pressure profiles albedo	Forcing Parameterization Validation
MIMR	Precipitation, snow cover, soil moisture albedo, clouds fraction, aerosols, soil moisture bidirectional reflectance distribution function	Forcing Parameterization Validation
MRIS	Temperature and water vapor profiles, cloud cover, albedo, surface temperature, snow cover, aerosols, surface resistance/ evapotranspiration land cover classification, vegetation indices, leaf area index/fractional Photo-synthetically Active Radiation	Parameterization Validation
SAGE III	Cloud height, H_2O concentration and mixing ratio, temperature and pressure profiles	Validation
TRMM	Rainfall profile, surface precipitation (precipitation radar, TRMM microwave imager, visible/infrared imaging radiometer	Forcing, validation
Other	4 dimensional data assimilation, including wind, temperature, and humidity profiles; momentum, energy, and precipitation fluxes	Forcing, validation

Source: Running et al. (1997).

for sensors under EOS, but very little is oriented toward hydrological programs. Why? Because terrestrial hydrological modeling, through remote sensing, will require a suite of satellites while the validation plans tend to be oriented on a single-sensor basis. Table 1 gives a list of EOS instruments that are needed for land surface hydrologic modeling.

In the terrestrial ecology community, more progress has been made in establishing a coordinated global set of terrestrial observations (Running et al., 1997). A Terrestrial Observation Panel (TOP) was established under the Global Climate Observing System (GCOS), a body established to provide the observations needed to meet the scientific requirements for monitoring, detecting, and predicting climate change. TOP is developing a five-tier terrestrial observation plan and implementation strategy in conjunction with the WCRP and the International Geosphere-Biosphere Programme.

The five-tier approach helps to establish the level of financial and scientific activity required for different types of validation efforts. The approach recognizes the necessary participation of large but temporary projects like FIFE, HAPEX, and BOREAS for certain validations and more permanent, geographically distributed facilities like national resource station networks for other global validation activities. This organizing vision is essential to produce globally consistent and representative validations of the full suite of land science products. The potential exists for a similar plan for terrestrial hydrology, but little has been accomplished to date. Thus, the recommendation: *There is a need to establish a global validation plan for terrestrial hydrology, and the plan should be an integral component of the GCOS.*

SUMMARY AND CONCLUSIONS

This paper has attempted to evaluate progress in observations and measurements since the publication of *Opportunities in the Hydrologic Sciences.* The progress is mixed: On the positive side, there is a substantial body of new measurements, thanks in part to WCRP-coordinated experiments; there is increased awareness and appreciation among hydrologists that measurements and observations are a critical component of developing new hydrologic theory and models; and data availability has increased due to significant advances in computer processing and data storage and transmission. On the negative side, the hydrology community has failed to establish the data needs for either operational hydrology and water management or scientific hydrology. Because of this, terrestrial hydrology has been the "stepchild" of climate field experiments, resulting in incomplete or inadequate measurements that prevent basic computations like determining water and energy balances over a range of scales. In addition, operational data collection remains under fiscal pressure, resulting in short-term savings at the expense of long-term benefits. Meanwhile, the hydrology community has yet to clearly articulate the operational data needs of the nation.

REFERENCES

Baeck, M. L., B. Bauer-Messmer, J. A. Smith, and W. Zhao. 1998. Heavy rainfall: Contrasting two concurrent Great Plains thunderstorms. Weather and Forecasting 12(4):785-798.

Bauer-Messmer, B., J. A. Smith, M. L. Baeck, and W. Zhao. 1997. Heavy rainfall: Contrasting two concurrent Great Plains thurderstorms. Weather and Forecasting 12(4):785-798.

Beljaars, A. C. M., P. Viterbo, M. Miller, and A. K. Betts. 1996. The anomalous rainfall over the USA during July 1993: sensitivity to land surface parameterization and soil moisture anomalies. Mon. Wea. Rev. 124:362-383.

Betts, A. K., J. H. Ball, and A. C. M. Beljaars. 1993. Comparison between the land surface response of the European Centre model and the FIFE-1987 data. Q. J. R. Meteorol. Soc. 119:975-1001.

Betts, A. K., J. H. Ball, A. C. M. Beljaars, M. Miller, and P. Viterbo. 1996. The land surface-atmosphere interaction: A review based on observational and global modeling perspectives. J. Geophys. Res. 101(D3):7209-7225.

Chen, F., K. Mitchell, J. Schaake, Y. Xue, H. Pan, V. Koren, Q. Duan, M. Ek, and A. Betts. 1996. Modeling of land surface evaporation by four schemes and comparisons with FIFE observations. J. Geophys. Res. 101(D3):7251-7268.

Duan, Q. Y., J. Schaake, and V. Koren. 1996. FIFE 1987 water budget analysis. J. Geophys. Res. 101(D3):7197-7207.

Famiglietti, J. S., and E. F. Wood. 1994a. Multi-scale modeling of spatially-variable water and energy balance processes. Water Resour. Res. 30(11):3061-3078.

Famiglietti, J. S., and E. F. Wood. 1994b. Application of multi-scale water and energy balance model on a tallgrass prairie. Water Resour. Res. 30(11):3079-3094.

Jarvis, P. G. 1976. The interpretations of the variations in leaf water potential and stomatal conductance found in canopies in the field. Roy. Soc. Lond. Philos. Trans. Ser. B 273:593-610.

Liang, X., D. P. Lettenmaier, E. F. Wood, and S. J. Burges. 1994. A simple hydrologically based model of land surface water and energy fluxes for general circulation models. J. Geophys. Res. 99(D7):14415-14428.

Mason, R., and T. H. Yorke. 1997. Streamflow Information for the Nation. Fact Sheet FS-006-97. Office of Surface Water, U.S. Geological Survey, Reston, Va.

National Research Council. 1991. Opportunities in the Hydrologic Sciences. Washington, D.C.: National Academy Press.

National Research Council. 1996. Assessment of the Hydrological and Hydrometeorological Operations and Services. Washington, D.C.: National Academy Press.

National Weather Service. 1992. Flash Floods and Floods: A Preparedness Guide. NOAA/PA 92050. Washington, D.C.: U.S. Government Printing Office.

Peters-Lidard, C., M. Zion, and E. F. Wood. 1997. A soil-vegetation-atmosphere transfer scheme for modeling spatially variable water and energy balance processes. J. Geophys. Res. 102(D4):4303-4324

Running, S., J. Collatz, J. Washburn, and S. Sorooshian (eds.). 1997. EOS Science Implementation Plan, Land Ecosystems and Hydrology. Washington, D.C.: National Aeronautics and Space Administration.

Sellers, P. J., F. G. Hall, G. Asrar, D. E. Strebel, and R. E. Murphy. 1990. The First ISLSCP Field Experiment (FIFE). Bull. Am. Meteorol. Soc. 71(10):1429-1447.

Sellers, P. J., F. G. Hall, G. Asrar, D. E. Strebel, and R. E. Murphy. 1993. An overview of the First International Satellite Land Surface Climatology Project (ISLSCP) Field Experiment (FIFE). J. Geophys. Res. 97(D17):18345-18371.

Shao, Y., and A. Henderson-Sellers. 1996. Modeling soil moisture: A project for inter-comparison of land surface parameterization schemes phase 2(b). J. Geophys. Res. 101(D3):7227-7250.

Smith, J. A., M. L. Baeck, M. Steiner, and A. J. Miller. 1996a. Catastrophic rainfall from an upslope thunderstorm in the Central Appalachians: The Rapidan Storm of June 27, 1995. Water Resour. Res. 32(10):3099-3113.

Smith, J. A., D. Seo, M. L. Baeck, and M. D. Hudlow. 1996b. An intercomparison study of NEXRAD precipitation estimates. Water Resour. Res. 32(7):2035-2045.

Viterbo, P., and A. Beljaars. 1995. An improved land surface parameterization scheme in the ECMWF model and its validation. J. Climate 8:2716-2748.

4

Ground Water Dating and Isotope Chemistry

Fred M. Phillips
Department of Earth and Environmental Science
New Mexico Institute of Mining and Technology

INTRODUCTION

Ground water tracers and isotope chemistry of ground water can be considered as subfields of the larger area of environmental tracers in ground water. Environmental tracers are simply chemical or isotopic solutes that are found in ground water as a result of ambient conditions rather than the deliberate activity of a researcher. They are studied mainly for the information they give about the ground water flow regime rather than the nature of the chemical activity in the ground water system. Such tracers have assumed new prominence in the past decade as a result of the refocusing of attention in applied ground water hydrology from questions of ground water supply, which are somewhat independent of the details of the flow path, to questions of ground water contamination, for which understanding the flow path and the nature of solute transport along it are central. *Opportunities in the Hydrologic Sciences* (NRC, 1991) emphasizes that "environmental isotopes are a key tool in studying the subsurface component of the hydrologic cycle."

Despite recently increased interest in applications of environmental tracers, no clear path of development over the past 5 to 10 years can be laid out. This diffuse and unpredictable nature of development is a direct outcome of the opportunistic nature of the field. Scientific disciplines that have a large theoretical component (e.g., the mathematical description of solute transport in ground water) tend to develop in a coherent way as individual researchers explore and build upon earlier theory, with only occasional correction from experimental results. New developments are driven in large part by intellectual assessment of immediately preceding work. In contrast, the application of environmental tracers to ground water hydrology has tended to be driven in large part by the introduction of analytical technologies developed by workers in other fields. Although in some cases the systematics of the tracer behavior have been worked out during investigations of ground water systems, more commonly the systematics have been previously well understood from independent investigations and the focus has mainly been on what the tracers can reveal about ground water flow and transport.

Instead of attempting to trace development of the field over the past few years, which might result in a relatively unenlightening catalog of methodologies, a few of the most innovative applications will be highlighted here. Some of the questions raised by *Opportunities in the Hydrologic Sciences* will then be addressed. That report focuses on the nature of scientific problems in hydrology and the means by which solutions to those problems might be efficiently promoted: "The choice of research problems is occasioned by its level of development within the hierarchy of the science, by the availability of new methods with which to solve it, and by the desire to understand a hydrologic phenomenon more deeply. The solution to the problem advances the development of the science and expands the conceptual framework that gives it meaning. . . That is the challenge to hydrologic science" (p. 7). Later the report states that "achieving this comprehensive understanding will require the kind of long-term disciplinary and interdisciplinary effort that can be sustained only by a vigorous scientific infrastructure" (p. 13) and, additionally, that "the supporting scientific infrastructure, including distinct educational programs, research grant programs, and research institutions, does not now exist for hydrologic science and must be put in place" (p. 2). Two research problems in environmental tracers will be examined in this paper, both with societally important and immediate applications but with very different histories. The goal is to elucidate how interactions with the "scientific infrastructure" affected the development of these problems and how characteristics particular to environmental tracer research influenced those interactions.

RECENT DEVELOPMENTS

Vadose Zone

One major trend in vadose zone hydrology has been a new interest in the behavior of water in arid-region vadose zones, mostly as a result of the need to predict contaminant transport at waste disposal sites. This has resulted in an increased emphasis on environmental tracer methods, partly because tracers are directly relevant to predicting the movement of dissolved contaminants and partly because the time scales for flow in arid vadose zones are often so slow that information from short-term physical monitoring may be difficult to extrapolate to the longer scale appropriate for solute transport.

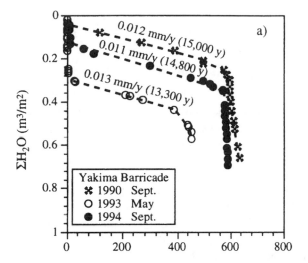

FIGURE 1   Cumulative water volume as a function of cumulative chloride mass (both per unit area) for three boreholes in the Pasco Basin, Washington. The recharge rates are calculated from the slope of the line and the total chloride accumulation times (in parentheses) from the chloride inventory. Based on the geological history of the site, the actual accumulation time is known to be between 13,000 and 15,000 years. Source: Murphy et al. (1996). © 1996 by American Geophysical Union.

One frequently used tracer in this situation is also one of the simplest—chloride. Chloride inventories with depth are commonly used to estimate net infiltration rates (see Figure 1), and increases in concentration are used to estimate evapotranspiration (Allison et al., 1994). Ancillary tracers include tritium (Scanlon, 1992), ^{36}Cl (Tyler et al., 1996), and the stable isotopes of hydrogen and oxygen (Liu et al., 1995). Although a shallow upper zone of arid-region soils is typically hydraulically active, response time in deep desert vadose zones is on the scale of 10^3 to over 10^4 years (Phillips, 1994; Murphy et al., 1996). Even the basics of water flow in arid-region vadose zones are still incompletely understood. This setting constitutes one of the frontiers of hydrology at the present time.

Shallow Aquifers

This setting is currently of great interest because the preponderance of ground water contamination problems are found there. For many years tritium was the tracer of choice. However, the inexorable radio decay of the pulse of tritium released into the environment by the atmospheric nuclear weapons test of the early 1960s has reduced tritium concentrations in most ground water to the point where they are of little use. Instead, attention has focused on the application of chlorofluorocarbons and combined $^3H/^3He$ to understanding shallow flow regimes. These methods are described in the following section.

Deep Aquifers and Regional Flow Systems

Probably the most exciting development in this setting has been the adaptation of solid source mass spectrometric methods originally developed for "hard rock" geochemistry to the investigation of heavy isotope ratios in deep ground water. At great depths the hydraulic properties are generally very poorly known and deep flow systems may as much reflect processes under ancient tectonic and climatic regimes as they do the influence of current conditions. In these circumstances tracers that can yield information on flow paths and rates are invaluable. Musgrove and Banner (1993) and Stueber et al. (1993) have demonstrated the utility of $^{87}Sr/^{86}Sr$, $^{147}Sm/^{144}Nd$, and other isotopes in unraveling the ancient flow regimes of the American midcontinent (see Figure 2), as have Moldovanyi et al. (1993) for the Gulf Coast Basin.

An area in which tracing studies in deep aquifers have contributed to the broader scientific framework is the increasing use of water in such aquifers (or minerals precipitated from the water) as archives of information about paleo-

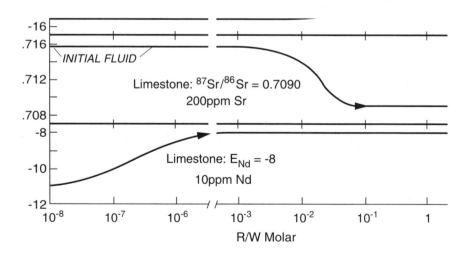

FIGURE 2   Calculated variations in three isotopic tracers as a function of progressive rock-water interaction (moles of rock dissolved and reprecipitated per mole of water) for ground water flowing through a limestone aquifer. The differing rates at which the water compositions respond to reaction with the rock suggest applications as tracers of water source and flow path history. Source: Reprinted, with permission, from Banner et al. (1989). © 1989 from Elsevier Science.

climatic conditions in the aquifer recharge area. The most notable examples are the studies by Stute et al. (1992, 1995) that used noble gas paleothermometry to establish the glacial-to-interglacial temperature reduction and that glacial temperatures were strongly reduced in the tropics. Other studies have focused on the reconstruction of variations in the stable isotope composition of precipitation over time (Winograd et al., 1992; Plummer, 1993; Clark et al., 1997). This is one of a relatively small number of instances where ground water science has made a significant impact on disciplines other than traditional subsurface fields.

CASE STUDIES IN THE DEVELOPMENT OF ENVIRONMENTAL TRACERS

The 1991 National Research Council report *Opportunities in the Hydrologic Sciences* played a major role in formalizing research support for hydrology as a separate discipline through establishing a separate National Science Foundation (NSF) program focused on hydrology. Ironically, this recognition was achieved just as internal pressure (i.e., the level of competition for research funding) rose to new highs and as the national rationale for funding scientific research was called into question as a result of ending the Cold War and of increased international economic competition. One reaction to the new circumstances was to call for increased emphasis on research that had relatively immediate societal applications. Hydrology clearly falls into this category; indeed, one common criticism of hydrological research is that the direction of research is driven too much by perceived applications. Nevertheless, no expansion of funding for research on hydrologic tracers has resulted from this change in emphasis.

Another reaction to the changing circumstances has been to reevaluate some of the basic assumptions that have served as a rationale for national funding of basic research for the past 40 years. Does "pure" or "curiosity-driven" research really contribute to the national welfare or is it largely a sideshow to research whose results are immediately applicable to societal problems? How are basic research results turned into products that will benefit society? How can the scientific research establishment promote the transfer of basic research results into applications?

Many recent writers on this topic have attempted to generalize answers to these questions. This paper will take a different approach and examine two case studies from environmental tracer research in hydrology. These cases raise their own questions because of widely contrasting histories of development. In one case a technique that was demonstrated as feasible and that had obvious application to both immediate societal problems of ground water contamination and fundamental problems of transport theory sat "on the shelf" for almost 15 years before it was put into practice, at which point it was widely hailed. In the other case a truly "off the wall" curiosity-driven research idea with no evident application to societal problems was funded and, even before publication of the results,

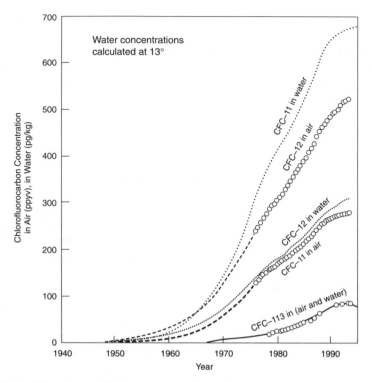

FIGURE 3   Histories of CFC concentrations in air and in water equilibrated with air, showing how measured concentrations in ground water can identify the date of recharge. Source: Szabo et al. (1996). © 1996 by the American Geophysical Union.

taken up and used to solve a major applied hydrology problem. Lessons can be learned from these examples, in the context of the fundamental nature of environmental tracer research.

Chlorofluorocarbons: Tracer Redivivus

Chlorofluorocarbons (CFCs) have many industrial uses, especially as working fluids in refrigeration units. Rates of atmospheric degradation are very slow. Thus, CFCs that have escaped after utilization have steadily accumulated in the atmosphere over the past 50 years (see Figure 3). They have only slight tendencies to adsorb on solid surfaces and are hence reasonably conservative in ground water. The key to their application as tracers is that, despite very low solubility in water, they are measurable at very low concentrations owing to their high electron affinity, which results in very sensitive responses in electron capture detectors attached to gas chromatographs.

Appreciation of the potential role of CFCs as ground water tracers was first developed in the early 1970s through collaboration between John Hays, a geochemist with a strong analytical orientation; Stanley N. Davis, a hydrogeologist; and Glen Thompson, a geology graduate student—all at the University of Indiana (Thompson et al., 1974). The work progressed, as Thompson's Ph.D. research, from the initial suggestion to methods development and limited field application (Thompson, 1976; Thompson and Hayes, 1979). Further development work continued until Thompson switched from ground water research to private consulting in the early 1980s. After that time virtually nothing was heard about CFCs as ground water tracers until the technique was revived by Niel Plummer and Eurybiades Busenberg of the U.S. Geological Survey (USGS) in the early 1990s (Busenberg and Plummer, 1991, 1992).

During this same period, the emphasis in ground water hydrology shifted from water supply to water contamination and solute transport. CFC tracing is now seen as a superb tool for understanding the shallow flow systems most susceptible to contamination. It can delineate recharge areas and rates, and flow directions and velocities, helping to establish sources and paths of contamination and predict future transport (Bohlke and Denver, 1995). During the 1980s, the focus of ground water theoretical development changed to solute (i.e., contaminant) transport, and elaborate tracer tests were conducted in order to test the theories. In retrospect, much of the same information could have been obtained by examining natural flow systems and applying CFC dating to provide a framework for the interpretation of the transport of environmental tracer inputs, such as the bomb tritium pulse (Szabo et al., 1996; see Figure 4). The widespread and rapid appreciation of the potential of the CFC method was demonstrated by the bestowal of the O. E. Meinzer Award to Plummer in 1993.

Why should a technique that was demonstrated as feasible, if not perfected, and was so immediately applicable to the burning questions of the day have been allowed to wither on the vine? Some of the answers to this question are clearly circumstantial. The departure of Thompson derailed the progress of the research. Bureaucratic factors also played a role. CFC samples are extremely subject to contamination and loss into or on container walls. Thompson overcame this difficulty by transporting his gas chromatographs into the field. However, the electron capture detector contained a small radioactive source, and during the 1980s regulatory controls on such sources virtually limited their use to laboratory settings. This problem was ultimately solved by Plummer through stringent sampling protocols and selection of container materials.

However, in addition to these fairly obvious obstacles, certain difficulties that are characteristic of environmental geochemistry research played a major role. The first of these is the issue of "art." "Art" refers to the range of skills and techniques that do not really fall within the domain of science but rather of craft. Geochemical research typically involves a great deal more art than many other fields in hydrology. Although the principles of a technique such as

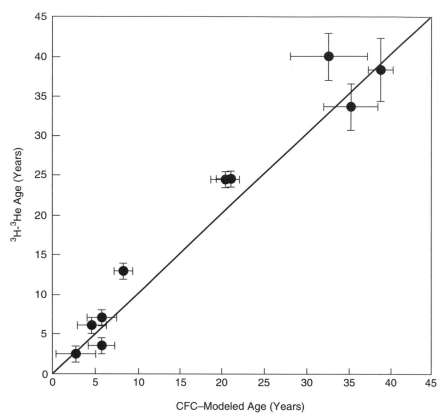

FIGURE 4   Comparison by Szabo et al. (1996) of the CFC and $^3H/^3He$ ages of shallow ground water in the Kirkwood-Cohansey aquifer system of southern New Jersey. The very good agreement between the two tracers indicates very minor dispersive mixing in the system. © 1996 by the American Geophysical Union.

analysis of very low levels of CFCs in natural water may be well understood, being able to reliably and reproducibly measure samples at a rapid enough rate for routine research applications may take years of tinkering with detector settings, column packings, valve assemblies and materials, carrier gases, data reduction programs, and literally hundreds of other experimental variables. Such art is rarely recorded; rather it is assimilated through apprenticeship. Any experimentalist knows that once such a technique is dropped, although the principles may be easily recovered, there will have to be an enormous investment of time to recover the art. This loss of expertise, in the case of CFC analysis, raised a barrier that forestalled most researchers who considered carrying on the method.

A second major impediment was presented by the nature of the research funding process. Developing the art as well as the science of the analytical method is time consuming; the (at that time standard) two-year grant required almost yearly new proposals, often before much demonstrable progress had been made. Without research funding programs directed specifically at hydrology, the proposals for continued funding had to compete with a very wide range of geochemical research projects. Geochemical analysis is instrumentation intensive, and some laboratories have succeeded in building up facilities worth millions of dollars. At least some of these laboratories took the viewpoint that the topics they researched were fundamentally so much more important than hydrological tracing that, in comparison, the new method did not deserve support. Expressed through the medium of proposal reviews, these opinions were successful in preventing continued funding for the research.

In this regard the USGS team led by Plummer had something of an advantage. Without being constrained by having to provide immediate results and without having to regularly run the gauntlet of hostile reviewers from other disciplines, they had the opportunity to systematically address the development of the method and present it to the hydrological community in a fairly mature state. The final factor that must also be mentioned is vision. Once a new geochemical method fails to reach fruition, even if this is due largely to circumstance rather than its merits, the huge investment of time and labor necessary to reinvent it raises a high barrier. It takes a researcher with a strong vision of the potential reward to take the risk of losing that investment.

Chlorine-36 in Fossil Rat Urine: Tracer Serendipitous

In the early 1990s I received a telephone call from an external reviewer for the U.S. Department of Energy plan of investigations at Yucca Mountain. The reviewer told me about the planned collection of ^{36}Cl samples from the exploratory shaft under the mountain. The basic idea was that samples would be collected at intervals as the shaft progressed and the $^{36}Cl/Cl$ ratio measured and compared with the modern atmospheric ratio, which is about 500×10^{-15}. Ratios higher than this value were considered indicative of very rapid infiltration of ^{36}Cl fallout from nuclear weapons testing in the 1950s. I had no connection with the Yucca Mountain project, but the reviewer had contacted me as an independent authority on hydrological applications of ^{36}Cl. When asked why she was so concerned about this particular aspect of the proposed investigations, she replied that it was considered the single most critical component of the entire range of research at the site, since the widespread presence of bomb ^{36}Cl would show that unsaturated fluxes were much greater than anticipated.

In fact, when the ^{36}Cl results came in (Fabryka-Martin et al., 1993), they showed that most of the ratios were above the modern atmospheric value (see Figure 5). However, instead of indicating that bomb ^{36}Cl was entering every-

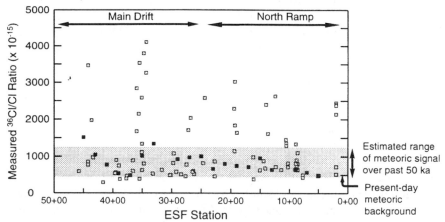

FIGURE 5 ^{36}Cl/Cl ratio as a function of distance (100-m increments) along the Exploratory Studies Facility (ESF) tunnel at the proposed high-level nuclear waste repository site, Yucca Mountain, Nevada. The present-day ^{36}Cl/Cl ratio is approximately 500×10^{-15}. Samples with ^{36}Cl/Cl higher than $1,250 \times 10^{-15}$ are thought to contain bomb ^{36}Cl. Source: Reprinted, with permission, Levy et al. (1997). © Los Alamos National Laboratory.

where, these were interpreted as indicating transport times in excess of 10,000 years (although it should be noted that some high ratios are still thought to indicate infiltration of fallout). Why the new interpretation? It was because a long-term history of much higher cosmogenic ^{36}Cl production had in the meantime been established through an entirely independent and quite unconventional research project.

In 1990 Edouard Bard had shown, by dating submarine corals using both U/Th and ^{14}C, that ^{14}C ages became systematically younger than the actual ages by several thousand years in the period from 12,000 to 20,000 years ago (Bard et al., 1990). This shift was attributed to increased ^{14}C production during that period, resulting from an increased cosmic-ray flux, which in turn was attributed to a weakened dipole geomagnetic field (Mazaud et al., 1991). However, this was difficult to demonstrate because ^{14}C activity in the atmosphere is also affected by shifts between the atmosphere/biosphere/marine reservoirs. After reading that paper, I realized that ^{36}Cl could potentially address the problem because it has a very short residence time in the atmosphere and essentially falls directly out onto the land surface. A record of ^{36}Cl deposition should therefore be much closer to a direct record of variations in cosmogenic production.

The difficulty was finding an archive in which ^{36}Cl deposition would be preserved, inasmuch as it is extremely mobile in the presence of water. Finally, I realized that fossil rat urine, preserved in ancient packrat middens of the southwestern United States (Betancourt et al., 1990), would preserve the chlorine that the rats ingested and excreted, in a context where the ^{36}Cl/Cl ratio could be assigned an age by means of ^{14}C dating of the urine itself. This experiment was proposed to the NSF and, after some initial difficulties, was funded. Midden samples from western and southern Nevada were obtained through the cooperation of Peter Wigand at the Desert Research Institute in Reno. The initial results were successful, and in a few years a sketchy preliminary history of ^{36}Cl deposition was reconstructed. The major feature of this reconstruction was a pattern of relatively low ratios during the Holocene (i.e., the past 10,000 years) compared to ratios higher by almost a factor of two prior to 13,000 years ago (Plummer et al., 1997; see Figure 6).

The significance of this result was immediately apparent to those involved in the Yucca Mountain project. Most of the unexpectedly high ^{36}Cl/Cl ratios beneath Yucca Mountain probably were a fingerprint of recharge prior to 10,000 years ago, rather than since 1950. The right information happened to appear just

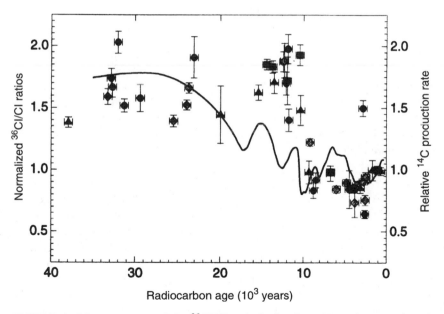

FIGURE 6   Measurements of the ^{36}Cl/Cl ratio in fossil packrat urine as a function of radiocarbon age. Chlorine-36 ratios are normalized to the modern ratio. The solid line indicates the history of variation in production of cosmogenic nuclides inferred from the record of variations in atmospheric radiocarbon activity. Source: Reprinted, with permission, from Plummer et al. (1997). © 1997 from the American Association for the Advancement of Science.

at the right time to prevent a major misinterpretation of some of the most critical data from one of the largest public works projects planned in the United States. Furthermore, it is obvious that this temporal variation in ^{36}Cl fallout can serve as a hydrological tracer in many other similar circumstances.

The contrast with the development of the CFC tracer method could hardly be more striking. Why was the secular variation ^{36}Cl tracer springing into existence just as it was needed, while the CFC tracer sat on the shelf for years after its capabilities had been highlighted? The most obvious answer has to do with research funding. Although the basic idea would undoubtedly have made a fine target for one of Senator Proxmire's "Golden Fleece" awards were he still alive, reviewers and program managers were able to see that it was a means to obtain critical data that could not be had otherwise. Other factors played a role. The "art" of ^{36}Cl analysis was maintained through long-term funding of the PRIME Lab accelerator mass spectrometry NSF facility at Purdue University and at my own laboratory, overcoming the inevitable start-up problems that stall the acquisition of real results.

REFLECTIONS

The state of research on environmental tracers in ground water hydrology is flourishing. New tracers are being developed, tracer results are being applied more and more widely as integral parts of hydrological investigations, and results of environmental tracer studies are being applied to fields outside hydrology. What can be done to promote continued growth of the field? Two case studies with remarkably contrasting histories were examined in order to reflect upon this question. Two studies cannot be presumed to yield any definitive patterns, but some common threads can be discerned:

1. The first point, and by far the most important, is a reaffirmation of the societal value of basic scientific research. This idea is certainly nothing new, but given the recent spate of attacks on this proposition, it is worth reemphasizing. Who could have predicted that the apparently straightforward and obviously applicable CFC tracer would not be successful for almost 20 years after it was first proposed, while the esoteric investigation of fossil rat urine would pay immediate practical dividends? The best criterion for predicting the utility of research directions is not an attempt to second-guess their end use, but rather some estimate of their fundamental scientific worth.

2. The long technique development times and the large investment in the geochemical art need to be recognized when supporting hydrological tracer research. Promising techniques may be killed by demanding immediate results. The trend toward longer NSF grant periods is a very promising development in this regard.

3. Proposed advances in environmental tracers need to be evaluated within

the context of the discipline of scientific hydrology. Possession of impeccable credentials in other areas of geochemistry does not qualify proposal reviewers from outside the field. The establishment of a separate NSF program in hydrology is a major step forward from the situation at the time CFC research temporarily died. For this step we owe a great deal to the *Opportunities in the Hydrologic Sciences* report and to those who invested their time in writing it.

REFERENCES

Allison, G. B., G. W. Gee, and S. W. Tyler. 1994. Vadose-zone techniques for estimating groundwater recharge in arid and semiarid regions. Soil Sci. Soc. Am. J. 58:6-14.

Banner, J. L., G. J. Wasserberg, P. F. Dubson, A. B. Carpenter, and C. H. Moore. 1989. Isotopic and Trace Element Constraints on the Origin and Evolution of Saline Ground Waters from Central Missouri. Geochim. Cosmochim. Acta 53:383-398.

Bard, E., B. Hamelin, R. G. Fairbanks, and A. Zindler. 1990. Calibration of the ^{14}C timescale over the past 30,000 years using mass spectrometric U-Th ages from Barbadoscorals. Nature 345:405-410.

Betancourt, J. L., T. P. Van Devender, and P. S. Martin. 1990. Packrat Middens: The Last 40,000 Years of Biotic Change. Tucson: University of Arizona Press.

Bohlke, J. K., and J. M. Denver. 1995. Combined use of groundwater dating, chemical, and isotopic analyses to resolve the history and fate of nitrate contamination in two agricultural watersheds, Atlantic coastal plain, Maryland. Water Resour. Res. 31:2319-2340.

Busenberg, E., and L. N. Plummer. 1991. Chlorofluorocarbons (CCl_3F and CCl_2F_2): Use as an Age Dating Tool and Hydrologic Tracer in Shallow Ground-Water Systems. U.S. Geological Circular 1992.

Busenberg, E., and L. N. Plummer. 1992. Use of chlorofluorocarbons (CCl_3F and CCl_2F_2) as hydrologic tracers and age-dating tools: The alluvium and terrace system of central Oklahoma. Water Resour. Res. 28:2257-2284.

Clark, J. F., M. Stute, P. Schlosser, S. Drinkard, and G. Bonani. 1997. A tracer study of the Floridan aquifer in southeastern Georgia: Implications for groundwater flow and paleoclimate. Water Resour. Res. 33:281-290.

Fabryka-Martin, J. T., S. J. Wightman, W. J. Murphy, M. P. Wickham, M. W. Caffee, G. J. Nimz, J. R. Southon, and P. Sharma. 1993. Distribution of chlorine-36 in the unsaturated zone at Yucca Mountain: An indicator of fast transport paths. Pp. 58-68 in Proceedings of FOCUS '93: Topical Meeting on Site Characteristics and Model Validation. September 26-29, 1993, Las Vegas, Nevada.. La Grange Park, Ill.: American Nuclear Society.

Levy, S. S., D. S. Sweetkind, J. T. Fabryka-Martin, P. R. Dixon, J. L. Roach, L. E. Wolfsberg, D. Elmore, and P. Sharma. 1997. Investigations of Structural Controls and Mineralogic Associations of Chlorine-36 Fast Pathways in the ESF. Report LA-EES-1-TIP-97-004. New Mexico: Los Alamos National Laboratory.

Liu, B., F. Phillips, S. Hoines, A. R. Campbell, and P. Sharma. 1995. Water movement in desert soil traced by hydrogen and oxygen isotopes, chloride, and chlorine-36, Southern Arizona. J. Hydrol. 168:91-110.

Mazaud, A., C. Laj, E. Bard, M. Arnold, and E. Tric. 1991. Geomagnetic field control of ^{14}C production over the last 80 ky: Implications for the radiocarbon time scale. Geophys. Res. Lett. 18:1885-1888.

Moldovanyi, E. P., L. M. Walter, and L. S. Land. 1993. Strontium, boron, oxygen, and hydrogen isotope geochemistry of brines from basal strata of the Gulf Coast sedimentary basin, USA. Geochim. Cosmochim. Acta 57:2083-2099.

Murphy, E. M., T. R. Ginn, and J. L. Phillips. 1996. Geochemical estimates of paleorecharge in the Pasco Basin: Evaluation of the chloride mass balance technique. Water Resour. Res. 32:2853-2868.

Musgrove, M. L., and J. L. Banner. 1993. Regional ground-water mixing and the origin of saline fluids: Midcontinent, United States. Science 259:1877-1882.

National Research Council. 1991. Opportunities in the Hydrologic Sciences. Washington, D.C.: National Academy Press.

Phillips, F. M. 1994. Environmental tracers for water movement in desert soils of the American Southwest. Soil Sci. Soc. Am. J. 58:15-24.

Plummer, L. N. 1993. Stable isotope enrichment in paleowaters of the Southeast Atlantic Coastal Plain, United States. Science 262:2016-2020.

Plummer, M. A., F. M. Phillips, J. Fabryka-Martin, H. J. Turin, P. E. Wigand, and P. Sharma. 1997. Chlorine-36 in fossil rat urine: An archive of cosmogenic nuclide deposition during the past 40,000 years. Science 277:538-541.

Scanlon, B. R. 1992. Evaluation of liquid and vapor water flow in desert soils based on chlorine-36 and tritium tracers and nonisothermal flow simulations. Water Resour. Res. 28:285-298.

Stueber, A. M., L. M. Walter, T. J. Huston, and P. Pushkar. 1993. Formation waters from Mississippian-Pennsylvanian reservoirs, Illinois Basin, U.S.A.: Chemical and isotopic constraints on evolution and migration. Geochim. Cosmochim. Acta 57:763-784.

Stute, M., M. Forster, H. Frischkorn, A. Serejo, J. F. Clark, P. Schlosser, W. S. Broecker, and G. Bonani. 1995. Cooling of tropical Brazil (5°C) during the last glacial maximum. Science 269:379-383.

Stute, M., P. Schlosser, J. F. Clark, and W. S. Broecker. 1992. Paleotemperatures in the southwestern United States derived from noble gases in ground water. Science 256:1000-1002.

Szabo, Z., D. E. Rice, L. N. Plummer, E. Busenberg, S. Drinkard, and P. Schlosser. 1996. Age dating of shallow groundwater with chlorofluorocarbons, tritium/helium 3, and flow path analysis, southern New Jersey coastal plain. Water Resour. Res. 32:1023-1038.

Thompson, G. M. 1976. Trichloromethane: A New Hydrologic Tool for Tracing and Dating Ground Water. Ph.D. thesis, University of Indiana.

Thompson, G. M., and J. M. Hayes. 1979. Trichlorofluoromethane in ground-water. A possible tracer and indicator of groundwater age. Water Resour. Res. 15:546-554.

Thompson, G. M., J. M. Hayes, and S. N. Davis. 1974. Fluorocarbon tracers in hydrology. Geophys. Res. Lett. 1:177-180.

Tyler, S. W., J. B. Chapman, S. H. Conrad, D. P. Hammermeister, D. O. Blout, J. J. Miller, M. J. Sully, and J. M. Ginanni. 1996. Soil-water flux in the southern Great Basin, United States: Temporal and spatial variations over the last 120,000 years. Water Resour. Res. 32:1481-1499.

Winograd, I. J., T. B. Coplen, J. M. Landwehr, A. C. Riggs, K. R. Ludwig, B. J. Szabo, P. T. Kolesar, and K. Revesz. 1992. Continuous 500,000-year climate record from vein calcite in Devils Hole, Nevada. Science 258:255-260.

5

Streamflow Prediction: Capabilities, Opportunities, and Challenges

Stephen J. Burges
Department of Civil Engineering
University of Washington

INTRODUCTION

The annual Wolman Lecture of the National Research Council's (NRC) Water Science and Technology Board is in honor of one of the great scientist-engineers of the century. This colloquium, which expands the scope of the annual lecture, is a fitting way to discuss some of the developments that resulted from the Eagleson committee's report *Opportunities in the Hydrologic Sciences* (NRC, 1991a). Peter Eagleson provided the intellectual drive and force for that committee. The report followed an earlier one by a committee chaired by Walter Langbein (Ad Hoc Committee on Hydrology, 1962). The Langbein committee's report followed an incisive assessment of "The Field, Scope, and Status of the Science of Hydrology," by Robert Horton (Horton, 1931). It is fitting that there are vignettes of Abel Wolman, Walter Langbein, and Robert Horton on pages 29, 44, and 41, respectively, of the 1991 NRC report.

The conclusions from the Eagleson committee's report included: "To meet emerging challenges to our environment we must devote more attention to the hydrologic science underlying water's geophysical and biogeochemical role in supporting life on earth. The needed understanding will be built from long-term, large-scale coordinated data sets and, in a departure from current practices, it will be founded on a multidisciplinary education emphasizing the basic sciences. The supporting educational and research infrastructure must be put in place" (NRC, 1991a, p. 11).

Sixty years earlier Robert Horton noted that

> A complete list of the problems of hydrology is impossible. It would be nearly co-terminous with a list of the applications of hydrology in both pure and ap-

plied science and in addition would involve much that is fundamental to the science itself. . . . As in physics and other sciences, an advance towards the solution of one problem uncovers others. The central problem is that of determining the physical processes and principles and the quantitative relations involved in the hydrologic cycle—or less comprehensively, as put by Ed. Imbeaux, the solution of the runoff problem. This problem serves also as an illustration of the fact that, in general, science can only progress as fast as the necessary quantitative data become available (Horton, 1931, p. 199)

We will see Horton's "uncovering of problems" demonstrated in several of the illustrations in this paper.

The chosen theme of this paper involves hydrologic applications to societally important issues associated with "runoff." All applications discussed depend on skills culled from many disciplines and are rooted in the basic sciences. The issues are concerned with effecting the water budget for catchments of various scales, with emphasis on floods and droughts. A comprehensive coverage of the hydroclimatology related to floods and droughts, flood forecasting and drought prediction, water supply forecasting, and flood and drought management, is given in Paulson et al. (1991). The challenges we face in predicting the streamflow for floods and droughts are exciting and daunting and will push us to the limits of intellectual and technological capabilities. The concerns that Horton had about availability of suitable data are equally relevant today.

STREAMFLOW PREDICTION

The most important input to the land surface for hydrologic predictions is precipitation. Inability to forecast, measure, or model the spatial and temporal amounts and form of precipitation will limit any analysis, interpretation of data, or attempts to model the hydrologic response to that precipitation. Hoyt et al. (1936) demonstrated the importance of determining areal rainfall accurately for flood estimation using unit hydrographs. Dawdy and Bergman (1969) provided one of the earliest comprehensive model demonstrations of the importance of rainfall variability and uncertainty for streamflow hydrograph predictions. In catchments where snowfall constitutes a major part of the incident precipitation, accurate depiction of the spatial coverage and depth of snow is essential to streamflow prediction. Many have investigated aspects of this problem. An illustration of the nature of the problem for complex terrain and possibilities for using spatially distributed modeling is given by Wigmosta et al. (1994). They used digital elevation data (180-m grid) to illustrate potential applications for a spatially distributed hydrology vegetation model that they developed to estimate water yield from the Middle Fork Flathead River in northwestern Montana. Satellite observations of snow cover were used to test the model's spatial predictive capabilities. Their work represents the state of the art in spatial modeling and emphasizes the need for more complete and informative spatial data for model inputs and environmental decision making.

Time and Spatial Scales—Droughts

There are several time and spatial scales of concern in drought prediction. Much depends on the definition of drought. If "agricultural drought" is the major issue, prediction of the spatial and temporal patterns of precipitation over relatively large areas is needed. The prediction time scales range from a month or so before scheduled planting, during the growing season, and near the end of the growing season. The timeliness of the prediction as well as its accuracy is of crucial economic importance to agribusiness. The early-season forecasts must be sufficiently accurate to influence seed planting and other farming decisions.

A second time scale involves multiple year below-critical precipitation patterns. This requires semiquantitative forecasts of large-scale atmospheric circulation patterns and associated hydroclimatological balances at the mesoscale to estimate the state of water distribution in the soil column throughout the area. The associated issue of precipitation recycling through regional reprecipitation of some of the evaporated water puts additional demands on the need to couple hydrological and meteorological models suitable for making hydroclimatic forecasts. Eltahir and Bras (1996) have provided a comprehensive review of the significance of precipitation recycling at continental and regional scales.

For much of the world where water is stored in reservoirs for later redistribution for societal needs, additional hydroclimatological forecasting is needed, particularly concerning streamflow. Good estimates are needed of the time patterns of streamflow to each reservoir as well as estimates of the release schedules necessary to meet societal contracts and environmental laws. The quality and timeliness required of the forecast depend on the size of the reservoir, the seasonal pattern of streamflow inputs, and the relative amount of water that is to be released to some schedule for societal and ecological purposes. For small reservoirs (capacity is a small fraction of the mean annual flow volume of the river and releases are also relatively small), short-term forecasts (on the order of weeks to several months) are all that are needed. The most important forecasts are for anticipated seasonal flow patterns.

Over-year storage is provided when the reservoir is on the order of the mean annual flow volume or larger and the annual release schedule is on the order of 50 percent or larger of the mean annual inflow volume. For much of the United States the combined demands placed on release schedules for reservoir systems put them into the over-year category. When reservoirs are relatively full, accurate short-term forecasts are needed for operational management purposes. When reservoir levels are lower, accurate long-term forecasts are essential. Some relief can be achieved with interties in complex systems. There is no relief, however, for complex systems that all experience the same broad regional climate and areally extensive shortage of streamflow. It is such forecasts that are hardest to make and of growing importance to society.

Time and Spatial Scales—Floods

The issue of floods poses a different set of requirements. There are multiple time and spatial scales. Burges (1989) addressed the issue of trends in forecasting and hydrologic modeling of hazardous floods. He addressed principally the issue of real-time flooding. There is another issue of clustering of major floods (Barros and Evans, 1997) that tend to occur over relatively short periods. Predicting such vulnerability is important when implementing engineering works in flood-prone areas. Clearly, if we had fair knowledge that a period of clustered major flooding was anticipated, work could be postponed or increased construction risk management measures would need to be taken. The usual assumption of quasi-stationarity of floods used in risk assessment for the design of temporary works to keep work areas dry during construction would invalidate standard economic risk decisions. An additional issue involves riverine navigation. If superflood clusters were predicted, alternative transportation plans could be considered, although the reallocation might be marginal if the principal mode of transportation is riverine (e.g., in the Mississippi River Basin).

If we restrict the discussion to riverine flooding (and do not concern ourselves here with major issues of coastal, estuarine, and lake shore flooding), real-time or relatively short lead-time forecasts of river flow rates and inundation levels can have different degrees of usefulness and precision. Forecasts can be made with greatest precision when the lead time is short or when an upstream stream gage measures the hydrograph that, after propagation downstream, creates the flood hazard. We are best able to estimate the changes in a hydrograph as it propagates downstream with little local inflow between an upstream location and the downstream location of concern. We have much poorer skills in attempting to estimate what the hydrograph will be at an upstream location when we have only an estimate of rain to be expected at some future time. Even when we have measured rain, our best predictions are associated with main channel flood flow routing. In short, most skill is associated with "how to route," and least skill is associated with "what or how much to route."

Riverine flooding can be generated by one or several concurrent mechanisms. Depending on the time of year and geographic location, riverine flooding may be caused by rainfall, snowmelt, rainfall and snowmelt combined, catchment thawing, or ice breakup, movement, damming, and ice dam breaching. Riverine flooding can also be caused by movement of hyperconcentrated sediment-laden flood waters. When engineered facilities are involved, the impact of naturally generated streamflow may be mitigated by storage in dams, containment by levees, diversion through floodways, or by inundating locations that have been made flood proof or designated for sacrificial flooding.

In almost all forecasting situations, estimates or measurements of precipitation are needed for input to some appropriate model. In most cases this means measurement and processing of spatial rainfall at time t and predicting spatial rainfall over the catchment at time $t + \Delta t$. Georgakakos and Kavvas (1987) provide an extensive review of all aspects of precipitation modeling, analysis, and prediction as well as several suggestions for research that should lead to improved prediction. They have hopes for stochastic precipitation extrapolation where precipitation models are coupled with remote (radar) and ground sensors. Schaake (1989) has demonstrated how quantitative stochastic precipitation forecasts for 4×4 km grids with time steps of about 15 minutes and up to 3 hours lead time, coupled with catchment geomorphology (representation of spatial locations of first-order basins and dominant channel links), are important for flash flood warning schemes where hydrologic response times are short. Foufoula-Georgiou and Krajewski (1995) report on developments since the earlier assessment of Georgakakos and Kavvas. Much work remains to be done on approaches for including forecasts of spatial precipitation patterns into appropriate hydrologic models and flood warning systems.

The Art and Science of Flood Forecasting

There is still art and science involved in making precipitation forecasts and in estimating spatial rainfall from modern radar measurements. Krzysztofowicz (1995) provided an extensive review and assessment of advances associated with flood forecast and warning systems. Krzysztofowicz' assessment is that only a small amount of information is added over the predictive skill for rainfall spatial coverage and depth using storm dynamics models (on the order of one hour of lead time for convective rain) beyond what can be achieved by estimating the storm trajectory based on radar scans alone.

We depend heavily on radar (and ground measured rain) for rainfall estimates for real-time flood forecasting. Krzysztofowicz (1995) brought to the attention of the community and summarized the report of a case study by Amburn and Fortin (1993), who reported on the June 5, 1991, storm over Osage County in Oklahoma. This case study emphasized the importance of human judgment in issuing a flood warning for convective storms. Two rain gages in the basin of interest and nine nearby gages were used to estimate basin average rainfall (1.44 in.), which, when combined with a rainfall runoff model, indicated that the maximum river stage would be below the bank full-level and would not pose a flood threat. Modern radar (WSR-88D) estimates suggest a basin average rain depth of about 5.2 in. This radar estimated that rainfall depth for the storm would cause the hydrologic model to predict a major flood. The rain gages in the basin did not represent the rain depth adequately, and the radar reflectivities were increased by the presence of hail. Subjective inputs and reports from field observers were used to estimate basin average rainfall of 2.8 in. This latter estimate, used with a model, predicted a flood, and a timely warning was issued 2.5 hours before flood stage was reached.

This example is important in several ways. A relatively simple hydrologic model was used rather than a more complex spatial model. The inputs to the model were "best subjective and objective" estimates. The purpose of timely warning was achieved. Much remains to be done to sharpen radar estimates of rainfall depth. The spatial coverage of storm patterns, however, is helpful for spatial models, and we need to be working toward implementing them. The critical caveat is that the process is far from automation.

EXTREME STORMS AND FLOODS

One major hydrometeorological issue is assessing the adequacy of existing emergency spillways or hydrological design of emergency spillways. Critical inputs include complete spatial and temporal descriptions of extreme storms and the associated complete flood hydrograph. Few data for these extreme situations are available. An opportunity for studying an extreme storm and the associated basin response occurred with an extraordinary flood in the Rapidan River (drainage area of 295 km^2) on June 27, 1995. The peak flood flow rate of 3,000 m^{3}s^{-1} (or 10.2 m^{3}s^{-1}km^{-2}) fell on the envelope curve of maximum flood discharge per unit area in the United States for rivers east of the Mississippi River. Radar reflectivity for the storm was recorded by the WSR-88D radar at Sterling, Virginia, located at an ideal distance of 80 to 100 km from the basin boundaries (Smith et al., 1996a). Smith et al. (1996b) have provided an extensive analysis of the storm and its temporal movement. The highest estimated rainfall depth within the basin exceeded 600 mm in 6 hours.

Smith et al. (1996b) used considerable skill to reconstruct the likely rainfall history for the basin. Forensic hydrometeorology was needed for the reconstruction activities. There were no operational standard rain gages located in the basin, and the stream gage recording equipment was destroyed when the river reached high stage. Fortunately, however, diligent hydrologists from the Virginia Department of Environmental Quality were present at the stream gage location, and they recorded the stage variation with a video camera. The hydrograph was reconstructed from the visual record.

The standard 4 × 4 km radar precipitation estimates provided by the National Weather Service (NWS) indicated a total rainfall over the catchment that was one-third of the measured runoff. Smith et al. (1996b) give a detailed explanation for this gross underestimation. One of the causes was the use of a standard, but inappropriate, U.S.-wide (Z-R) relationship (derived for completely different raindrop distribution circumstances) between radar reflectivity (Z) and rainfall rate (R). Nonstandard rain gages provided approximate point estimates of total rainfall at five locations. These data, together with the limits imposed by the measured hydrograph, provided a basis for deriving a more realistic Z-R relationship.

In addition to this calibration of the radar reflectivities, Smith et al. (1996b) developed additional rainfall rate products at 6-minute intervals at the scale of 1 × 1 km to take advantage of all radar information. They also made extensive use of the radar volume scans to determine an atmospheric mass balance for purposes of estimating storm efficiency. The atmospheric sounding at Sterling, Virginia, provided additional crucial information for the analysis.

Some of the key findings from Smith et al. (1996b) are displayed here in Figures 1, 2, and 3. Figure 1 shows the reconstructed (estimated) discharge hydrograph and the radar-estimated rainfall input rate over the entire catchment. Figure 2 shows the centroid of the storm track. The dominant precipitation hugged the topographic ridges, demonstrating the considerable influence of local orography on storm movement and rainfall delivery. Figure 3 shows a map of radar-estimated total storm depth.

Smith et al. (1996b, p. 3105) observed that "a striking feature of the Rapidan storm was its small size and long duration. Rain area ranged from a minimum of approximately 50 km^2 to a maximum area of approximately 350 km^2." Three basins, "South," "Conway," and "Rapidan" are shown in Figure 2. The maximum aggregated discharge per unit area of all three subbasins (295 km^2) was 10.17 m^{3}s^{-1}km^{-2}. The discharge rates in the higher rainfall Conway and main stem Rapidan basins (Figure 3) would likely have been higher. How much higher is untestable.

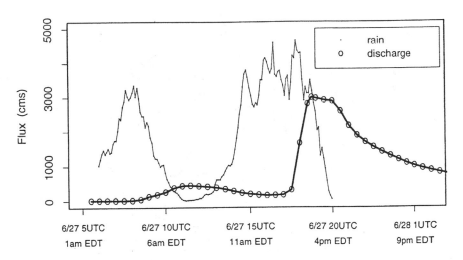

FIGURE 1 Estimated discharge hydrograph for the Rapidan River at Ruckersville, Virginia, and 6-minute time series of basin-averaged rainfall rate for the Rapidan River upstream of Ruckersville for June 27, 1995. Source: Reprinted, with permission, from Smith et al. (1996b, Fig. 3). © 1996 from the American Geophysical Union.

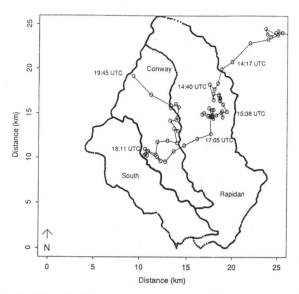

FIGURE 2   Track of the Rapidan storm as represented in surface rainfall centroid locations.   Source: Reprinted, with permission, from Smith et al. (1996b, Fig. 9). © 1996 from the American Geophysical Union.

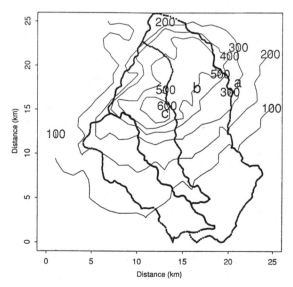

FIGURE 3   Map of storm total rainfall accumulations (in millimeters), June 27, 0000-2000 UTC (Universal Time Coordinates) at 1 km grid scale derived from Sterling (radar) volume scan reflectivity observations.   Source: Reprinted, with permission, from Smith et al. (1996b, Fig. 10). © 1996 from the American Geophysical Union.

This was an extreme storm. When the final analysis was completed, Smith et al. (1996b) estimated that the average rainfall over the 295-km^2 basin was 0.344 m and that runoff depth was 0.296 m. They also pointed out the critical importance of the Doppler velocity observations for estimating the atmospheric water budget for this storm. Their final observation has significant implications for engineering design:

> The influence of small-scale topographic features on Rapidan rainfall suggests that probabilities of catastrophic rainfall are locally variable and may be guided by the spatial characteristics of the watershed itself. This in turn implies that probabilities of mass wasting and flood impacts may be site-specific, even within a small area that otherwise appears to be climatically homogeneous . . . .

The influence of topographic features on Rapidan rainfall has significant implications for engineering hydrometeorology procedures used for design of high hazard structures, in particular Probable Maximum Precipitation (PMP) procedures. A cornerstone of PMP analysis is the storm transposition procedure. Transposition of the Rapidan storm to any other location is implausible. This brings into question the practice of transposing storms like the August 19-20, 1969 Virginia storm that exhibited strong links to topographic features but for which details of storm structure and evolution are not available (Smith et al., 1996b, p. 3112).

Rapidan Basin—Possibilities for Spatial Modeling of Runoff Production?

There are several questions concerning possibilities for spatial precipitation runoff modeling for situations similar to the Rapidan River basin. The basin has high relief and should be amenable to distributed spatial hillslope hydrologic modeling using models similar to that of Wigmosta et al. (1994). There was considerable mass wasting in the upper basins, so the assumed geometry and soil properties at the start of the storm changed appreciably during the storm. It is unclear how to model the hydrologic response of such a basin during extreme rainfall. The normal problem is determining the supply of water to the channels ("what" or "how much" to route). In the case of the Rapidan the available water would be difficult to determine in space and time. A large fraction became streamflow, so the larger questions in this setting would be "how to route water across and through changing hillslopes" and "how to route water, sediment, and other debris" through the changing channel system.

Policy Issues —Extreme Storm Data

There are significant policy issues related to the Rapidan storm. Smith et al. (1996b) have demonstrated the considerable utility of the full Doppler information obtainable from the U.S. NWS's WSR-88D radar reflectivity information. It is evident that effort needs to be placed by NWS personnel on developing accurate rainfall rate and storm rainfall depth products at scales useful to the hydro-meteorology community. This suggests shorter time and smaller spatial increments being developed and archived for use. The need for early and accurate calibration of radars is obvious. The importance of atmospheric sounding data is clear. More of such data is needed rather than less. Finally, the issue of how to calibrate radars must be addressed. Much has been written about the inadequacy of point rainfall for hydrologic modeling. It is clear to this observer that there is a critical need for a network of ground-based rainfall-measuring devices as well as disdrometer information if we are to make best use of the capability of radars. The kind of measurements needed is addressed below. There is also need for ground-based gages to supplement radar measurements where the radars miss the rain production parts of clouds.

ISSUES IN RADAR AND RAIN GAGE CALIBRATION

There are few locations where the rate and depth of rainfall reaching the ground can be estimated and measured by radar, a standard tipping bucket rain gage, and an electromechanical device for measuring drop-size distribution (disdrometer). Steiner et al. (1997) report on such measurements and estimation of rainfall at a climatological station located in Goodwin Creek near Oxford, Mississippi. The Goodwin Creek equipment is maintained by staff members of the National Sedimentation Laboratory of the U.S. Department of Agriculture's Agricultural Research Service (USDA-ARS). The tipping bucket gage is calibrated regularly; the ground-level disdrometer has been in place since April 1996 and is located approximately 2 m from the tipping bucket gage. The WSR-88D radar used to estimate rainfall over the Goodwin Creek site is located in Memphis, Tennessee, 121.2 km from the site. This is close to the optimum distance for radar rainfall estimation reported by Smith et al. (1996a). The ground-based instrumentation is to provide information suitable for calibrating the radar for use in estimating rainfall rate, rainfall amount, and rainfall kinetic energy. The objective is to provide detailed spatial and temporal patterns of rainfall kinetic energy for use with modern erosion estimation models.

The accumulated rainfall depths from eight storms were presented by Steiner et al. (1997). The recorded depth in the disdrometer and the corresponding percentage recorded by the standard tipping bucket gage ranged from a low of 61 percent to a high of 94 percent. Storm depths (mm) and the associated percentage caught by the tipping bucket gage from three representative storms were 16.9 and 71 percent, 5.8 and 61 percent, and 30.9 and 94 percent. There is no apparent simple correction scheme that can be implemented to adjust the tipping bucket rainfall rate record to reflect actual rainfall reaching the ground.

Figure 4 presents information for the storm of June 9, 1996, to emphasize variations in precipitation rate and accumulated depth as estimated from radar reflectivities or as measured directly. The instantaneous rainfall rate (averaged

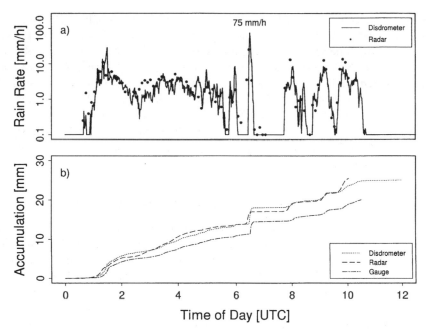

FIGURE 4   Time series of instantaneous rainfall rate determined from radar reflectivity and ground-based disdrometer and accumulated rainfall determined from radar reflectivity, disdrometer accumulations, and a tipping bucket rain gage, Goodwin Creek, Mississippi, June 9, 1996. Source: Reprinted, with permission, from Steiner et al. (1997). © 1997 from the University of Mississippi.

over 5 to 6 minutes) was estimated for a radar pixel of plan dimension 1×2.1 km ($1°$ azimuth at a distance of 121 km) corresponding to the lowest radar sweep at an angle of $0.5°$. The disdrometer measurements are recorded at 1-minute increments. Steiner et al. (1997) reported details of this storm as follows:

Rainfall was continuous from 0100 to 0600 UTC. Afterwards more of a cellular pattern was exhibited until rain stopped around 1100 UTC. Accompanied with a five-degree temperature drop from $21°$ to $16°$ Celsius within the hour following 0100 UTC, the relative humidity increased from around 90 percent to saturation. Wind speeds (10m elevation) were in the range of 3 to 5 m/s with a gust reaching 7 m/s. The following 10 hours, however, were very stationary with winds from the WNW blowing generally less than 3 m/s, the temperature remained at about $16°$ Celsius (though after 0800 UTC it decreased by another degree), and the relative humidity stayed around 100 percent.

The accumulated rainfall measured by the disdrometer was 25.1 mm. The corresponding amount measured by the tipping bucket gage was 20.1 mm, which was 80.4 percent of the calibrated ground-level disdrometer value. The peak (1-minute) rainfall rate recorded by the disdrometer was about 75 mm/hour. Figure 4a shows the time series of the radar-estimated rainfall rate and the rainfall rate measured with the disdrometer on a logarithmic scale. Figure 4b shows the accumulated rainfall depth on a linear scale. For this storm the radar-estimated rainfall and the disdrometer-measured rainfall rates and accumulated amounts are in close agreement. The accumulated catches from the disdrometer and the tipping bucket rain gage (there was no wind shield for the latter) diverged up to about 0006 UTC. There were catch differences associated with each rainfall spike for the remainder of the storm.

Policy Issues—Rain Gage and Radar Calibration

The different estimates of precipitation for the June 9, 1996, storm shown here highlight the need for much more rigorous analysis of precipitation as it reaches the ground than we have done in the past. The network of installed rain gages is likely to be used to help calibrate the NEXRAD network of radars in the United States. Steiner et al. (1997) have shown that careful examination of rainfall data recorded by the existing network of gages will be required before the data can be used to determine appropriate radar reflectivity-rainfall relationships. Disdrometer measurements as well as readings from rate-recording rain gages that are recessed such that their rims are at ground level ("buried gages") will be needed to help calibrate the network of radars. This will provide an opportunity for a comprehensive new ground-level rain-measuring network to be established.

CATCHMENT MASS BALANCE AT A RANGE OF SCALES

The fundamental problem of hydrology is determining the water balance for a specified time and region. The balance can be effected by using sufficient measurements of precipitation, runoff, ground water discharge, changes in soil water storage, and evaporation flux, if they are available. Usually a combination of measured data and model-interpolated or -estimated fluxes or states, typically for evaporation, soil moisture storage states, and ground water flow and storage, are used. Evaporation from land surfaces averages about 60 percent of precipitation (Brutsaert, 1986). While few direct measurements are made of evaporation, considerable work has been done in the past decade to describe and quantify turbulent transfer mechanisms in the atmospheric boundary layer to improve methods for parameterizing evaporation and transpiration at catchment and regional scales. Much of that work has been along the lines described by Brutsaert (1986); an excellent summary is given by Parlange et al. (1995).

The difficulty in effecting the water balance is knowing the spatial and temporal distributions of precipitation inputs and whether the precipitation is in liquid or solid form, as well as the spatial and temporal patterns of evaporation and transpiration. An added complication is associated with determining movement of water to the ground water zone and thereafter the movement of ground water at scales ranging from a few tens of meters of travel to a channel to hundreds of kilometers as regional ground water movement. Few measurement systems are established that track ground water movement adequately enough to permit a complete and accurate description of the water balance.

At relatively small scales (a few hectares), it ought to be possible to represent the relevant processes fairly accurately. Modeling and measurement efforts at small scale for two situations are discussed below to indicate some of the issues that are involved in improving streamflow prediction. The first illustration is for combining modeling and monitoring to represent the hydrology of small humid-zone catchments before and after land use change. The second is for representing the hydrology at small scale in a semiarid catchment where Horton (infiltration limited) overland flow is the primary mechanism by which water from the hillslope reaches the stream channel system. There are many other examples that could be used, but the two chosen demonstrate the range of issues of interest.

Small-Scale Catchment Hydrology—Humid
Mediterranean-Marine Climate

Wigmosta and Burges (1997) have presented an adaptive hydrologic monitoring and modeling approach designed to describe the surface hydrology of forested catchments in the humid Pacific Northwest. The objective was to use the minimum monitoring consistent with a spatially distributed model of the dominant hydrologic processes (1) to describe the spatial wetting and drying patterns throughout a catchment during the year, (2) to predict the surface outflow, and (3) to estimate evapotranspiration and recharge to underlying aquifers. The model needed sufficient spatial resolution to be useful in describing the same features after the land had been converted to suburban use.

Figure 5 shows schematically the processes that are represented. Evaporation and transpiration are represented as physical water transport rather than as energy fluxes. Any desired degree of spatial disaggregation can be used for precipitation input and representation of hillslope and channel hydrologic processes. The acronyms HOF and SOF correspond to Horton Overland Flow and Saturated Overland Flow, respectively. Modeling and measurement are done continuously in time. Several years of continuous in-time (typically at a time increment of 15 minutes) concurrent modeling and measurement are needed to calibrate the model. Two example applications were reported, one involving a 37-ha forested catchment where monitored data for two wet seasons were used to calibrate the model, and a 16.7-ha suburban catchment where data for one water year were used for calibration. In both cases the model was used to estimate

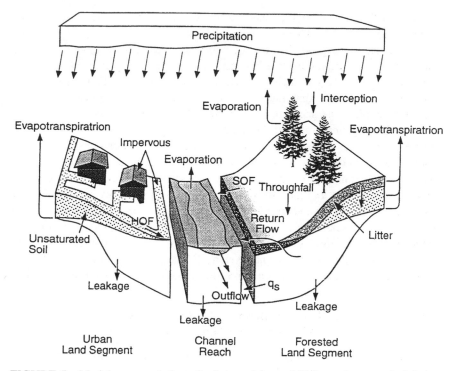

FIGURE 5   Model representation of urban and forest hillslope elements draining to a channel reach. Each hillslope element is modeled as a vertical two-dimensional section aligned with the dominant downhill flow path. Source: Reprinted, with permission, from Wigmosta and Burges (1997, Fig. 1). © 1997 from the American Geophysical Union.

catchment areally averaged evapotranspiration and leakage through the underlying till layer (aquifer recharge). A single rain gage was used in both situations to represent precipitation input.

   Annual summary information for the four water years for the forested catchment are shown to illustrate aspects of the catchment mass balance. Figure 6 shows the principal wet and dry season distributions of water for the 37-ha forested catchment. All quantities are expressed as depth of water averaged over the catchment. The forest soil thickness ranged from about 0.8 to 1 m and was underlain by dense till. Precipitation and runoff depths were measured and evapotranspiration (ET) and till recharge were estimated by the model. There are several key observations. Precipitation is distinctly seasonal. The wet (October 1 to April 30) and dry (May 1 to September 30) season classifications are arbitrary but correspond to the dominant rain and lesser rain periods. If we think of the catchment (37 ha or about 91 football fields) as a pixel in a larger-scale situation, we could represent the inputs and outputs as part of a distributed representation of the hydrology of the landscape.

Flow production from the catchment is small to negligible in the dry season, but more than 50 percent of the evapotranspiration occurs then. The two years where wet and dry season evapotranspiration rates were almost equal correspond to low wet season rainfall with increased evaporation occurring during the wet season. The till leakage provides recharge to a deeper aquifer that provides sustained water supply to down-basin streams. ET is not known accurately, so a small change in the real ET could significantly alter the actual recharge. The need for accurate estimation of the residual leakage for aquifer health and stream ecology is obvious. It is equally obvious that it is a nontrivial exercise and that hydroclimatic data need to be recorded for a range of weather patterns to provide guidance for effecting the relevant water budget using a combination of modeling and hydrologic measurements. At a minimum, some shallow piezometers (about

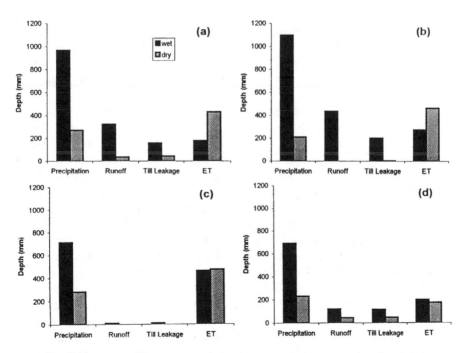

FIGURE 6   Wet (October 1 to April 30) and dry (May 1 to September 30) season measured precipitation and runoff and calculated till leakage and evapotranspiration for water years (a) 1990, (b) 1991, (c) 1992, and (d) 1993, for a 37-ha forested catchment. Source: Reprinted, with permission, from Burges et al. (1998, Fig. 7). © 1998 from the American Society of Civil Engineers.

1 m deep) are needed to provide critical information about variations in soil water storage.

The implications for land use change when the forest soil is removed and compacted for urban development ought to be obvious. Most municipality stormwater managers are concerned with the runoff part of the balance. With a thinner soil there will be less recharge in this setting and a change in ET and runoff patterns. The ecological importance of recharge is often overlooked.

The illustration presented above is for a situation where continuous measurement and modeling are required to effect the water balance. In the following example, Horton runoff production from individual storms is the principal hydrologic issue of concern.

Small-Scale Catchment Hydrology—Semiarid Monsoon Climate

Goodrich et al. (1995) conducted careful measurements of convective rainfall on an intensively monitored experimental catchment, Lucky Hills—104, which is a part of the USDA-ARS Walnut Gulch Experimental Watershed. The measurements were made during the summer monsoon season (July-September) of 1990. Lucky Hills—104 catchment has an area of 4.4 ha (about nine football fields). The hope when representing the hydrology of such a relatively small catchment with low topographic relief is that a single rain gage would be suitable for effecting the hydrologic balance.

Figure 7 shows the contours of total rainfall for the storm of August 12, 1990. Measurements at 48 nonrecording (cumulative) rain gages were used to construct the rain contour map; the maximum contour shows a depth of 55 mm. While the rainfall cumulative total is relatively uniform, a single tipping bucket recording rain gage is inadequate to capture all the features of the spatial rainfall pattern that are needed to predict flow production and delivery to the channel.

Figure 8 shows the modeled and measured hydrologic responses for the 4.4-ha catchment for the short-duration storm of August 3, 1990. The average rainfall depth was approximately 12.7 mm, and the measured runoff volume was 3.5 mm peak were insensitive to Manning's n on averaged over the catchment. This figure shows the end result of using the state-of-the-art distributed Horton runoff model of the USDA-ARS-KINEROSR, the research version of KINEROS (Woolhiser et al., 1990). The best-modeled hydrographs are shown when using five combinations of one, two, three, and four recording rain gages to estimate the spatial rainfall coverage of this 4.4-ha catchment. This figure emphasizes the need for more than one rain gage in an extremely small catchment and the importance of the spatial locations of those gages if the response to convective rainfall is to be modeled accurately. (The finest-resolution radar rainfall pixel is 1×1 km, which is 22.7 times larger than the Lucky Hills—104 experimental catchment.) This gives pause when considering forecasting flood response from convective storms in larger catchments in semiarid areas.

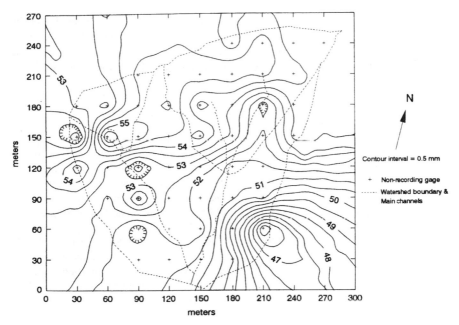

FIGURE 7   Contour map of the rainfall depth for the storm of August 12, 1990, at Lucky Hills (interpolation by isotropic Kriging).   Source: Reprinted, with permission, from Goodrich et al. (1995, Fig. 5).   © 1995 from Elsevier Science.

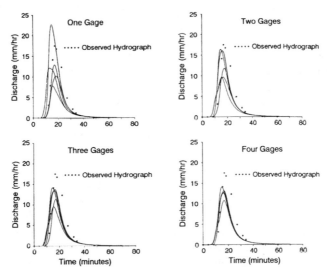

FIGURE 8   Modeled hydrographs for five combinations of one, two, three, and four recording rain gages in Watershed LH-104 (August 3, 1990).   Source: Reprinted, with permission, from Faures et al. (1995, Fig. 2).   © 1995 from Elsevier Science.

Woolhiser (1996) presents a comprehensive discussion of physically based rainfall-runoff models used to estimate hydrographs in principally Hortonian runoff production situations and their strengths, weaknesses, and potential. He emphasizes the importance of knowledge of spatial and temporal rain patterns:

> One cannot disagree with the fact that it is difficult or impossible to calibrate models with many interacting parameters. . . . The works of Goodrich (1990) and Michaud and Sorooshian (1994) give us valuable insight into problems of calibration for watersheds when Hortonian runoff is the primary runoff generation mechanism, and where infiltration into the channels becomes more important with increasing basin scale. It is especially revealing that parameter sensitivities change dramatically with both basin scale and the magnitude of the rainfall input. For example, Goodrich (1990) found that runoff volume, peak rate, and time to peak were insensitive to Manning's n on both planes and channels for watershed LH-106 (0.36 ha), while Michaud and Sorooshian (1994) found that runoff characteristics were very sensitive to Manning's n in channels . . . for WG-1 (150,000 ha). . . . The importance of spatial and temporal resolution of rainfall data cannot be overemphasized (Woolhiser, 1996 p. 126).

Large-Scale Catchment Hydrology

The two small catchment examples that have been given are for the situation where the control volume is the watershed and the inputs are defined by measured (or spatially estimated) precipitation at the ground surface. As the area of interest increases, it is important to couple hydrology with atmospheric water and energy balances. In principle, if the control volume were extended above the watershed to include the complete airshed and all water and energy fluxes across both boundaries were measured, it would be effective to do approximate major catchment plan view pixel hydrologic modeling using the moisture accounting models developed 30 years ago. The need to model surface energy fluxes as well as hydrologic fluxes and water storage states, at whatever pixel size is being used, has necessitated the development of models that do both energy and moisture accounting. A model that contains essential subsurface hillslope representation of appropriate flow physics, evaporation, transpiration, snow accumulation, melt, and ablation has been developed and demonstrated by Wigmosta et al. (1994). The model is intended primarily for use in complex terrain, and its use was demonstrated at a pixel scale of 180 m. In principle, it should be able to be coupled with a mesoscale meteorology (MM) model, which in turn should be nested within a general circulation model (GCM). The application demonstrated by Wigmosta et al. (1994) did not included this coupling. This is an area for cooperative multidisciplinary research and application.

Simpler conceptual vertical water storage models that release water laterally to streams and vertically to the atmosphere (and in some case to ground water recharge) have been developed for use at relatively large pixel scales in continental river-basin-scale hydrologic modeling compatible with the scale of GCMs. Liang et al. (1994) developed a two-layer variable infiltration capacity model (VIC-2L) that conceptualizes statistically the spatial distribution of soil moisture and runoff production zones within a catchment. The model has been tested against moisture and sensible and latent heat flux data collected during the First SLSCP Field Experiment (FIFE) experiment in Kansas. The model has been applied by Abdulla (1995) and Abdulla and Lettenmaier (1997) to the Arkansas-White-Red River Basin using a $1^\circ \times 1^\circ$ (approximately 10,000 km^2) land surface grid size. The VIC-2L model was calibrated to catchments ranging in size from 100 to 10,000 km^2 to permit application to the basin using regionally estimated parameters.

This form of modeling effort has pushed the limit of data handling and has required use of relatively coarse-scale data sets of soil classification, vegetation cover, etc. The work has shown a direction that is needed for the future. It represents what is needed for the GEWEX Continential-Scale International Project (GCIP) component of Global Energy and Water Cycle Experiment (GEWEX). Abdulla's work and that of workers who are continuing his efforts point to the need for more coordinated data sets of regional streamflow time series and more complete attempts to define basin precipitation coverage.

The bulk of the work to date involving hydrology at the continental river basin scale has focused on coupling hydrology with GCM and mesoscale meteorology representations of the surface water and energy fluxes and corresponding river flow. The enormously complicated issue of determining the (residual) recharge of aquifers and the importance of regional-scale groundwater movement has received limited attention. While the pioneering efforts of the early part of the GCIP program are commendable, much more remains to be done in the context of the complete water balance.

ISSUES IN LONGER-TERM STREAMFLOW PREDICTION

So far this discussion has been concerned with relatively short-term forecasts of flood flow given estimates of precipitation patterns and issues of effecting the water balance starting with measured precipitation. There are many issues, principally the influence that spatial and temporal patterns of water have on society, where predictions of future water availability are extremely important. All of these predictions are concerned with climatic variability, whether it will be generally wetter or drier than usual for time spans on the order of months to seasons to years to many years. The important hydrologic challenge is how to convert broad and uncertain predictions of climatic variability into corresponding patterns of channel flow, and spatial patterns of evaporation and transpiration, soil moisture states, and ground water recharge. In an ideal setting the predictions would be done using coupled GCMs, MMs, and hydrologic models to ensure that

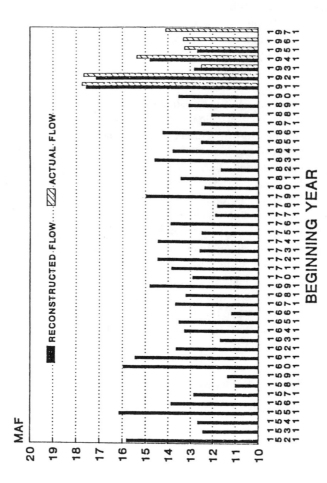

FIGURE 9 Bar plot of nonoverlapping 10-year means for reconstructed and actual flows at Lee's Ferry, Colorado River. "Actual data" are corrected "virgin flows" provided by the U.S. Bureau of Reclamation. Source: NRC (1991b).

the climatic predictions are compatible with precipitation recycling and ground surface (and vegetation) thermal fluxes.

We are aware of the forms of climatic variability from studies of historical information. A "drought atlas" is available on CD-ROM for the conterminous United States (Teqnical Services Inc., 1997) that has summaries of precipitation and streamflow probabilities for individual stations and adjacent regions. It can be used to get a sense of the risk to drought that the society in a given region faces. Probabilities are provided, starting in any given month, for the likelihood that rainfall will be above particular depths for time increments up to five years. Similar information is available for streamflow at specific locations. This product is available to the community largely through the efforts of James R. Wallis, a former member of the Water Science and Technology Board.

Analysis of longer-term streamflow records and paleosurrogates of streamflow volumes, most notably tree ring growth indices, has shown that sustained periods of above- and below-"normal" patterns are not uncommon. Figure 9 shows decadal average reconstructed streamflow volumes for the Colorado River at Lee's Ferry. This figure shows clearly that there are decadal periods of substantially above- and below-average river flow volume.

Figure 10 shows systemwide river flow volume for the Salt-Verde River Basin in Arizona. There is a significant range in the annual inflow volume as well as distinct long-term patterns of substantially below-average flow volume.

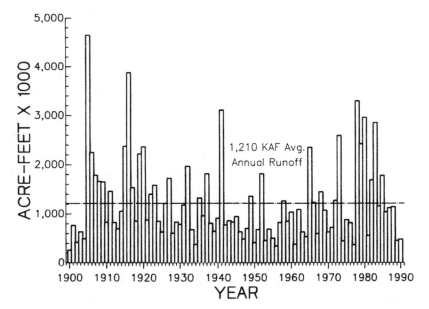

FIGURE 10   Annual runoff in the Salt, Tonto, and Verde river watersheds between 1900 and 1990. Source: NRC (1991b).

In this instance water supply is managed by increased pumping from aquifer reservoirs in dry years and recharging during wet years. Conditional probabilistic information, given the current states of river flow and reservoir storage, concerning the likelihood of low-, medium-, or high-flow conditions during the next several years, would be beneficial for developing operation strategies for such systems.

The historical record gives us other information that suggests actions that are needed for extended climatic variability prediction. Figure 11 shows an apparent relationship between the grouping of superfloods in the Upper Mississippi and Missouri river basins and aspects of the Southern Oscillation Index (SOI) between 1955 and 1985. A superflood was defined as a flood from a basin having an area greater than 10,000 km^2 whose peak response was thought to exceed the "100-year" magnitude. Figure 11 suggests a close link to the negative phase of SOI during El Niño events. El Niño events last on the order of two to seven years. Relatively long lead forecasts of the nature and extent of El Niño events would provide important hydroclimatic policy information for the region.

Predictability of "Low-Frequency" Climatic Variability

Superflood groupings and spatially and temporally extensive low-flow conditions are associated with broad features of global atmospheric circulation. Barros and Evans (1997) indicate that the spatial distributions of sea surface temperatures (SSTs), the cryosphere, and atmospheric aerosols influence the dominant atmospheric moisture circulation patterns and the resulting climatologies, weather patterns, and corresponding hydrology. The Holy Grail of the field is to couple complete ocean thermodynamics with the atmosphere and hydrosphere. Eagleson (1986) provides a clarion call to the community with his paper "The Emergence of Global-Scale Hydrology" in which he stresses the importance of global-scale modeling of the coupled ocean-atmosphere-land surface to address issues of hydrologic persistence. Much work has been done since then and much remains to be done. In a recent report, Gu and Philander (1997) show how interdecadal climate fluctuations depend on oceanic exchange between the tropics and extratropics. This emphasizes the importance of pursuing in various ways the Holy Grail. So far, GCMs have been operated with SST patterns placed in known locations and also with varying assumptions concerning the state and spatial and temporal patterns of the cryosphere. The field is at an early stage of development.

Most-low frequency relationships between broad continental U.S. climate and weather patterns have been tied in some way or another to El Niño-Southern Oscillation (ENSO) anomalies. Considerable efforts have been expended to determine related forecast skill, with particular interest in estimating the spatial and temporal patterns of SST anomalies. Ropelewski and Halpert (1996) have attempted to quantify measured Southern Oscillation precipitation relationships

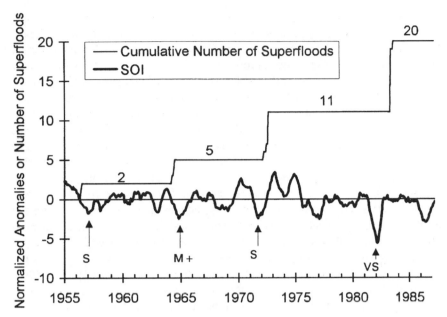

FIGURE 11   Monthly anomalies of Southern Oscillation Index and number of superfloods for the Upper Mississippi and Missouri river basin. Arrows indicate strength of El Niño-Southern Oscillation Events as defined by Quinn et al. (1987). VS = very strong, S = strong, M+ = moderate/strong. Source: Reprinted, with permission, adapted from Barros and Evans (1997, Fig. 1). © 1997 from the American Society of Civil Engineers.

principally in tropical areas. The objective was to provide long-range forecasters with quantitative guidance when making seasonal and multiseasonal predictions. They note, however, that there is considerable variation in the spatial variations of patterns in precipitation percentiles in some regions. Cane et al. (1997) report on trends in the twentieth-century SST patterns and suggest causes that include global warming. Ji et al. (1996) report on the state of the art of forecasting skill of the National Centers for Environmental Prediction (NCEP) coupled general circulation models. Numerous measures of forecast skill could be used. Based on Ji et al.'s Figure 11 (model CMP10), anomaly correlations indicate that for the ENSOs in the 1982-1992 period the model had valuable forecast skill for about 6 months and some useful skill for up to 12 months beyond. For the 1992 to 1995 period the skill level dropped to about 4 months. Even with these limitations, using historical weather patterns in a "fuzzy" forecasting mode might have some use. What is not clear is whether CMP10 predictions would be useful for predicting superflood conditions sufficiently far in advance to be of broad benefit to society.

While the search for physical cause-effect relationships for making long lead-time forecasts of broad atmospheric patterns is important and may offer the best hope for future directions, pragmatic forecasts are still needed and a variety of approaches are in use. One approach that appears to have useful forecasting skill has been developed by Lall and Mann (1995) and Lall et al. (1996). Lall and Mann (1995) used Singular Spectral Analysis and Multitaper Spectral Analysis to identify high fractional variance bands in the time series of climatic variables in the Great Salt Lake region as well as volume changes for the lake. The frequency bands of interest were 15 to 18, 10 to 12, 3 to 7, and 2 years. The interannual variations were consistent with ENSO signals and suggested that there may be forecasting predictability for this relatively large-scale closed basin.

Lall et al. (1996) treat the biweekly Great Salt Lake volume time series (1847 to 1992) as the output from a finite-dimensional nonlinear dynamical system and used nonlinear regression to discern the apparent dynamics. The resulting models (Multivariate Adaptive Regression Spline—MARS) were used for forecasting and showed substantial success. The ability to make forecasts for periods between 1.2 to 4.16 years ahead was demonstrated for the period 1984 to 1994. Figure 12 demonstrates this apparent forecast success. Given the observations of Ji et al. (1996), it is not known how well the scheme would have worked for the shorter-duration ENSOs after 1992. Lall et al. (1996, Figure 6) also shows successful four-years-ahead forecasts. The closed Great Salt Lake Basin is a highly persistent long-memory hydrologic system. The apparent success that Lall et al. (1996) have had suggests that there may be benefits from trying the approach in other highly persistent systems, particularly those that are correlated in some way to persistent ENSO-like signals.

Probabilistic Forecast and Extended Streamflow Prediction

Krzysztofowicz (1995) describes the various Quantitative Precipitation Forecast (QPF) products that were being produced by the then Techniques Development Laboratory and the National Meteorological Center (now NCEP) of the U.S. National Weather Service. More recently, a National Research Council committee (NRC, 1996) has emphasized the need for a closer connection between QPFs and how they will be used in hydrologic models for hydrologic prediction, particularly short-term flood warning. The report provides guidance for the modernization efforts of the NWS. In recommendation 3-10 of the report, the committee recommended that "the Office of Hydrology . . . should consider spatially distributed, continuous simulation hydrology models to replace/augment spatially lumped and parametric models. . . ."

An important part of the NWS modernization is the work of the Advanced Hydrologic Prediction System (AHPS) for water resources management. The Water Resources Forecasting System (WARFS) has development of long lead-time forecasts of streamflow as a key goal. It is clear that this is an important

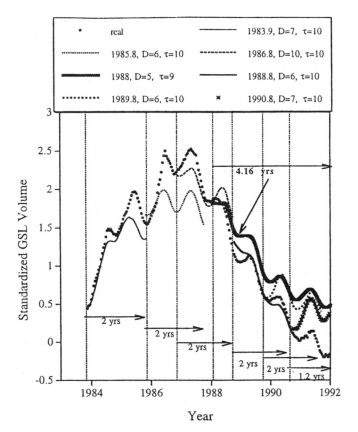

FIGURE 12   Forecasts of 1.2 to 4.16 years from different starting points during the 1983 to 1992 rise and fall of the Great Salt Lake, Utah. Source: Reprinted, with permission, from Lall et al. (1996, Fig. 8). © 1996 from the American Geophysical Union.

national priority, reiterated in recommendation 3-18 (NRC, 1996): "Field personnel and users of products and services should have a greater involvement in the further definition and development of the WARFS and other components of AHPS." AHPS will be undergoing a demonstration testing starting in March 1997 (Carlowicz, 1996) using the Des Moines River in Iowa as a test case. The target is to provide 25, 50, and 75 percent exceedance flow rate forecasts for days, to weeks, to multiple months ahead.

The potential benefits from spatial modeling and implementation of the WARFS program highlighted here are important. Resource prioritization is likely to be an issue with reductions in federal government funding. If a priority had to be established, work toward spatial representation of hydrologic processes (spatial modeling) to take advantage of the spatially variable precipitation input (provided by radar and rain gage networks) in regions subject to flash floods is appropriate and timely. An all-pervasive need is long lead-time forecasts of water supply in locations of the country where resource use is already high.

A related issue is the information needs and contributions of GEWEX. Cahine (1997) makes the case for additional effort to explore skill capabilities in seasonal and interannual climate prediction. This will require more rather than less effort from the modernized NWS. Atlas (1997) summarizes the state of the modernization program and emphasizes successes that have been achieved. Gains in 6- to 12-months-ahead forecasts depend on the full implementation of the Automated Weather Information Processing System. Much remains to be done to realize the full potential of modernization of the NWS.

We are in the relatively early stages of making hydrologic use of probabilistic QPFs. Few hydrologists have determined how to make use of them, which suggests a greater need for those making the forecasts and those using them to work closely. The older extended streamflow predictions issued by river forecast centers of the NWS have been used by water resource managers. Crowley (1993) has reported on a modeling scheme to provide probabilistic forecasts of basin runoff, over-lake precipitation, lake evaporation, net basin supply, and over-lake air temperature for the Great Lakes Basin. He illustrates the approach with a 6-months-ahead probabilistic forecast of five broad classes of net basin supply for Lake Superior. The range of potential conditions could readily be put into context with the experience for the previous autumn-winter period. Crowley (1996) describes an approach that uses all existing historical information as well as the long lead-time probabilistic QPF. This permits building a large set of possible future hydrologic spatial time series from which outlook probabilities and other parameters can be estimated. An example shows the probabilistic net basin supply forecasts by month for one year ahead. Crowley's work is indicative of how the work of atmospheric scientists and hydrologists can be combined to yield societally useful products for resource management decision making.

A Final Grand Challenge

Figure 13 shows the time series of May to April water year cumulative inflow volume to the surface reservoirs that serve the southwestern region of western Australia, including the Perth metropolitan area, for the period from 1911 to 1994. The annual inflow volume is expressed in billions of liters (giga liters, or GL). A giga liter can be visualized as the volume occupied by a 1-m depth of water covering a square kilometer. There are many notable features of this figure.

First, the range of inflow volume from year to year is large, with three years exceeding 800 GL and four years less than 100 GL. The reservoirs currently are targeted to supply approximately 180 GL. The average inflow volume for the

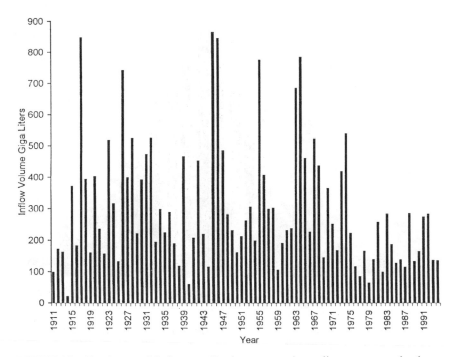

FIGURE 13   Total annual inflow to Perth, western Australia, water supply dams for May to April water years 1911 to 1994.

entire period is approximately 320 GL. For the period 1975 to 1994 it is 187 GL. The desired supply fraction to the mean inflow level for these two periods is 56 and 96 percent, respectively. The reservoirs are of sufficient size and the nature of the inflow variability for the entire record is such that they are adequate for supplying 56 percent of the long-term mean annual flow. They are considerably undersized to supply 96 percent of the recent mean annual flow for the last 20 years of the record.

The region of interest extends approximately 120 km north-south and 60 km east-west. The catchments are all located east of the Darling scarp, which is approximately 20 km inland from the coastline. There is a general reduction of rainfall and streamflow from the south to the north. The uncertainty of inflow volumes to the multiple reservoir system is a major cause of concern for the society. The population served is approximately 1.2 million. The population to be supplied by the system of surface and ground water sources is anticipated to grow to 1.7 million by the year 2011.

Determining if the recent 20 years is a likely guide to the future is a non-trivial task. To illustrate the demands that this situation places on hydrologic

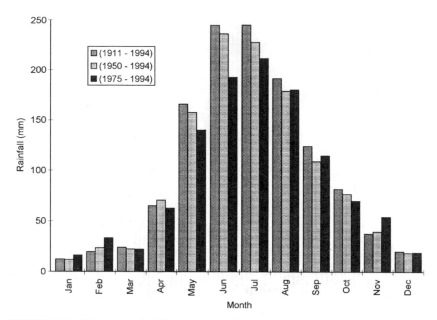

FIGURE 14   Monthly rainfall (mm) near Serpentine Dam, a part of Perth, western Australia, water supply system.

science, consider summary features of the seasonal rain supply and associated inflow to one of the reservoirs, Serpentine Dam, in the southeastern part of the supply area. The monthly average rainfall near Serpentine Dam is shown in Figure 14. (The annual average rainfall is 1,232 mm, and the average for the 1975-1994 period is 1,061 mm.) The bulk of the rain falls in five months during the winter. There has been a reduction in the mean monthly rainfall for May, June, and July in the past 20 years relative to the complete record. The corresponding reduction in inflow is dramatic and is shown in Figure 15. June and July inflow are each about one-half of the long-term average; there are noticeable reductions in August, September, and October. The reduction in rainfall and inflow to Helena Reservoir, approximately 45 km to the north, is more noticeable. The average annual rainfall is 1,045 mm and for the 1975-1994 period 865 mm. The reductions in inflow were dramatic. Inflow in June and July was about a third of the long-term average; in August, September, and October it was about 40 percent of the long-term average.

The rain that falls on the Perth water supply catchments infiltrates readily into the highly permeable lateritic soils. If a substantial storm follows closely a series of antecedent storms that have made the soil column relatively wet, subsurface flow is delivered to the channels. For storms with short interarrival times there is more surface flow generated than for storms having longer interarrival times, all other conditions being equal. Consequently, aggregated measured (or estimated) monthly rainfall depth information that does not contain complete information about the individual storms (depth, duration, and time between storms) has extremely limited value for streamflow prediction. Most longer-term estimation schemes do not yet contain this necessary detail.

Water supply systems operators and planners need multiyear forecasts of rainfall amounts for planning and operating their systems. Long-term forecasts are needed for the capacity expansion problem to determine the most dependable mix of supply between ground water, which requires expensive treatment, and water from surface reservoirs. Surface reservoirs fill in years of large positive excursions in the inflow volume from the long-term mean. This means it is important to be able to estimate the relative frequency of years when inflow will be larger than normal.

Figure 15 shows the relative deficit of large excursions in the past 20 years. Larger inflow volumes are associated with larger than usual wet season rainfall. Prediction of likely increased wet season rainfall should be a target. For wet

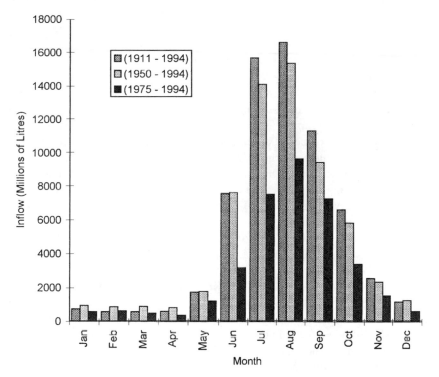

FIGURE 15 Monthly inflow in millions of liters to Serpentine Dam, Perth water supply system.

season rainfall closer to the average or lower than average, prediction of likely temporal storm patterns is necessary, particularly concerning the timing for large storms following other large storms. This poses a "grand challenge" for coupled ocean and climate models, particularly for nested mesoscale meteorology models. Refined prediction of inflows to surface reservoirs will place heavy demands on improved rainfall-runoff models that can track accurately the apportionment of water between recharge to ground water and movement to streams by predominantly subsurface flow paths.

Burges (1996) summarized the findings of research needs identified by research scientists and engineers who participated in a workshop that was held in Perth in 1996 to address the critical water supply issue for the growing Perth region. The summary calls for pushing the scientific limits of coupled ocean, atmospheric, and hydrologic models particularly the limits of prediction of likely precipitation over decade-long periods.

SUMMARY AND RECOMMENDATIONS

The examples presented emphasize the need for hydrologists to have a broad-based education and to be able to work closely with, and pose critical questions to, colleagues in ocean and atmospheric sciences, civil and environmental engineering, ecology, water systems management, and emergency preparedness personnel. There is a need for both generalists and specialists. The problems involved are multidisciplinary. All the examples involve various aspects of the water balance for a catchment. Two issues come to the fore. The first is the need to predict precipitation inputs in space and time for time horizons ranging from minutes to multiple years. We must learn to make use of even fuzzy long-term predictions for environmental management and associated societal decision making. The second issue concerns the need to improve the measurement or estimation of precipitation at a range of spatial scales. Calibration of weather radars should be given high priority. The hopes that accurate spatial precipitation coverage and depth will be provided in the near future, mainly by radar measurement, are perhaps optimistic.

There is considerable need for a new network of accurate ground-based measuring systems to augment what will be provided from radar coverage. This is evident from recent work by Kuligowski and Barros (1996) and the above-cited works by Smith et al. (1996b), Krzysztofowicz (1995), Goodrich et al. (1995), and Steiner et al. (1997). Much is likely to be achieved from modernization of the NWS. We all must work cooperatively to ensure that the NWS takes a leadership role in developing techniques and delivering products that are beneficial to society. There is a great deal to be gained by the NWS supporting external research groups and individuals to help achieve its modernization goals.

Much remains to be done in hillslope hydrology, as indicated by the work of Goodrich et al. (1995) and Burges et al. (1998). Renewed effort is required to develop hydroclimatology schemes that are compatible with features of spatially distributed hydrologic models and associated measurement systems. Measurement systems will combine ground-based measurement and radar estimation of precipitation, features of boundary layer water and energy transport, and remotely sensed atmospheric and ground-level hydrologic states and fluxes.

In all that is attempted there are many intellectual and developmental challenges. The "grand challenge" is working toward connecting ocean, atmosphere, and hydrosphere interactions into a coherent approach that will yield hydrologically useful information at the hillslope, catchment, and continental river basin scale for time scales up to the order of a decade. Mesoscale meteorology models and associated supporting measurements will provide the needed information at time scales on the order of minutes, hours, or several days. The need for longer-term information will require greater coordination and championing of the cause of research scientists and practitioners from hydrology, meteorology, and oceanography than has been done.

Scientific and managerial leadership, and coordination between programs in various agencies, has been provided by the founding program director for hydrologic sciences at the National Science Foundation, L. Douglas James, as he has worked to implement the findings of the Eagleson committee's report. Increased coordination is needed for the integration of hydrologic science with the atmospheric and ocean sciences to approach the many practical issues of the water budget and how it influences society and how society influences it. The Water Science and Technology Board has done a commendable job in covering the spectrum of water issues. The time appears to be ripe to establish a new NRC board to focus on hydrologic science, in its broadest context, including human-influenced ecosystems and societal infrastructure. Such a board would be charged with the development and nurturing of hydrologic science, particularly as it relates to water and society, and have as its primary charge to attend to all aspects of science associated with the hydrologic cycle.

The Eagleson committee's report was visionary and echoed leadership from an earlier era. Horton concluded his 1931 paper by saying that "the most immediate needs for the advance of the science are (a) the collection of additional basic data along various lines, (b) correlative research and coordination of existing results, and (c) research to provide connective tissue between related problems" (p. 202).

It is up to our generation to ensure that the leadership provided initially by the WSTB and the Eagleson committee is continued and that Robert Horton's clarity of vision is not lost. Horton's "connective tissue" is more important than ever.

EPILOGUE

We are clearly behind the times considering hydroclimatology and the timing of major climatic and storm features. Lerner and Loewe in Camelot reported on the state of meteorology and climatology at the time of King Arthur at Camelot:

ARTHUR: It's true! It's true! The crown has made it clear.
The climate must be perfect all the year.

A law was made a distant moon ago here:
July and August cannot be too hot.
And there's a legal limit to the snow here
In Camelot.
The winter is forbidden till December
And exits March the second on the dot.
By order, summer lingers through September
In Camelot.
Camelot! Camelot!
I know it sounds a bit bizarre,
But in Camelot, Camelot
That's how conditions are.
The rain may never fall till after sundown.
By eight, the morning fog must disappear.
In short, there's simply not
A more congenial spot
For happily-ever-aftering than here
In Camelot.

Boring as it would be to live in such a climate, a critical component is missing: the information is not quantitative!

ACKNOWLEDGMENT

The work reported here was supported in part with funds from the National
Science Foundation under grant EAR-9506391.

REFERENCES

Abdulla, F. 1995. Regionalization of a Macroscale Hydrological Model. Water Resources Series,
      Technical Report #134. Department of Civil Engineering, University of Washington.
Abdulla, F., and D. P. Lettenmaier. 1997. Application of regional parameter estimation schemes to
      simulate the water balance of a large continental river. J. Hydrol. 197:258-285.
Ad Hoc Panel on Hydrology. 1962. Scientific Hydrology. Washington, D.C.: U.S. Federal Council
      for Science and Technology.
Amburn, S. A., and S. Fortin. 1993. Use of WSR-88D and Surface Rain Gage Network in Issuing
      Flash Flood Warnings and Main Stem Flood Forecasts Over Osage County, Oklahoma, June 5,
      1991. Pp. 321-330 in NOAA Technical Memorandum NWS ER-87, Post-print Volume, Third
      National Heavy Precipitation Workshop. Washington, D.C.: U.S. Department of Commerce.
Atlas, D. 1997. Budgetary foul weather. Science 275:1719.
Barros, A. P., and J. L. Evans. 1997. Designing for climate variability. J. Prof. Issues Eng. Educ.
      Practice 123(2):62-65.
Brutsaert, W. 1986. Catchment-scale evaporation and the atmospheric boundary layer. Water
      Resour. Res. 22(9):39S-45S.

Burges, S. J. 1989. Hazardous floods: Trends in forecasting and hydrologic modeling. Pp. 210-228 in Proceedings of the Pacific International Seminar on Water Resources Systems, Tomamu, Japan.

Burges, S. J. 1996. Climate variability and water resources in the south-west of western Australia: Research needs and priorities. Pp. 21-23 in Climate Variability and Water Resources Workshop, Water and Rivers Commission (WA). J. K. Ruprecht, B. C. Bates, and R. A. Stokes, eds. Water Resources Technical Report Series No. WRT5. Pp. 21-23.

Burges, S. J., M. S. Wigmosta, and J. M. Meena. 1998. Hydrologic effects of land-use change in a zero-order catchment. J. Hydrol. Eng. 3(2):86-91.

Cahine, M. T. 1997. The success of seasonal-to-interannual climate prediction requires GEWEX. Global Energy and Water Cycle Experiment, World Climate Research Programme News 7(1):2.

Cane, M. A., A. C. Clement, A. Kaplan, Y. Kushnir, D. Pozdnyakov, R. Seager, S. E. Zebiak, and R. Murtugudde. 1997. Twentieth-century sea surface temperature trends. Science 275:957-960.

Carlowicz, M. 1996. NWS hydrologists predict better river forecasts. EOS Tran. Am. Geophys. Union 77(52):529-530.

Crowley II, T. E. 1993. Probabilistic great lakes hydrology outlooks. Water Resour. Bull. 29(5):741-753.

Crowley II, T. E. 1996. Using NOAA's new climate outlooks in operational hydrology. J. Hydrol. Eng. 1(3):93-102.

Dawdy, D. R., and J. M. Bergman. 1969. Effect of rainfall variability on streamflow simulation. Water Resour. Res. 5(5):958-966.

Eagleson, P. S. 1986. The emergence of global scale hydrology. Water Resour. Res. 22(9):6S-14S.

Eltahir, E. A. B., and R. L. Bras. 1996. Precipitation recycling. Rev. Geophys. 34(3):367-378.

Faures, J.-M., D. C. Goodrich, D. A. Woolhiser, and S. Sorooshian. 1995. Impact of small-scale spatial rainfall variability on runoff modeling. J. Hydrol. 173:309-326.

Foufoula-Georgiou, E., and W. Krajewski. 1995. Recent advances in rainfall modeling, estimation, and forecasting. Pp. 1125-1137 in U.S. National Report to International Union of Geodesy and Geophysics 1991-1994, Reviews of Geophysics, Supplement. Washington, D.C.: American Geophysical Union.

Georgakakos, K. P., and M. L. Kavvas. 1987. Precipitation analysis, modeling, and prediction in hydrology. Rev. Geophys. 25(2):163-178.

Goodrich, D. C. 1990. Geometric Simplification of a Distributed Rainfall-Runoff Model over a Range of Basin Scales. Ph.D. dissertation, Department of Hydrology and Water Resources, Tucson, University of Arizona. Tucson.

Goodrich, D. C., J.-M. Faures, D. A. Woolhiser, L. J. Lane, and S. Sorooshian. 1995. Measurement and analysis of small-scale convective storm rainfall variability. J. Hydrol. 173:283-308.

Gu, D., and S. G. H. Philander. 1997. Interdecadal climate fluctuations that depend on exchanges between the tropics and extratropics. Science 275:805-807.

Horton, R. E. 1931. The field, scope, and status of the science of hydrology. Trans. Am. Geophys. Union. Pp. 189-202.

Hoyt, W. G. 1936. Studies of Relations of Rainfall and Run-off in the United States. U.S. Geological Survey Water Supply Paper 772. Washington, D.C.: U.S. Government Printing Office.

Ji, M., A. Leetma, and V. E. Kousky. 1996. Coupled model prediction of ENSO during the 1980s and 1990s at the National Centers for Environmental Prediction. J. Climate 9(12):3105-3120.

Krzysztofowicz, R. 1995. Recent advances associated with flood forecast and warning systems. Pp. 1139-1147 U.S. National Report to International Union of Geodesy and Geophysics 1991-1994, Reviews of Geophysics, Supplement. Washington, D. C.: American Geophysical Union.

Kuligowski, R. J., and A. P. Barros. 1996. Evidence of orographic precipitation effects in the Appalachian Mountains. EOS Trans. Am. Geophys. Union 77(17):S112.

Lall, U., and M. Mann. 1995. The Great Salt Lake: A barometer of low-frequency climatic variability. Water Resour. Res. 31(10):2503-2515.

Lall, U., T. Sangoyomi, and H. D. I. Abarbanel. 1996. Nonlinear dynamics of the Great Salt Lake: Nonparametric short-term forecasting. Water Resour. Res. 32(4):975-985.

Liang, X., D. P. Lettenmaier, E. F. Wood, and S. J. Burges. 1994. A simple hydrologically based model of land surface water and energy fluxes for general circulation models. J. Geophys. Res. 99(D7):14,415-14,428.

Michaud, J., and S. Sorooshian. 1994. Comparison of simple versus complex distributed runoff models on a midsized semiarid watershed. Water Resour. Res. 30(3):593-605.

National Research Council. 1991a. Opportunities in the Hydrologic Sciences. Washington, D.C.: National Academy Press.

National Research Council. 1991b. Managing Water Resources in the West Under Conditions of Climate Uncertainty: A Proceedings. Washington, D.C. National Academy Press.

National Research Council. 1996. Assessment of Hydrologic and Hydrometeorological Operations and Services. Washington, D.C.: National Academy Press.

Parlange, M. B., W. E. Eichinger, and J. D. Albertson. 1995. Regional scale evaporation and the atmospheric boundary layer. Rev. Geophys. 33(1):99-124.

Paulson, R. W., E. B. Chase, R. S. Roberts, and D. W. Moody (Compilers). 1991. National Water Summary 1988-1989—Hydrologic Floods and Droughts. Water Supply Paper 2374. U.S. Geological Survey, Denver, Co.

Quinn, W. H., V. T. Neal, and S. E. A. Mayolo. 1987. El Niño occurrences over the past four and a half centuries. J.Geophys. Res. 92(C13):14449-14461.

Ropelewski, C. F., and M. S. Halpert. 1996. Quantifying southern oscillation—precipitation relationships. J. Climate 9(5):1043-1059.

Schaake, J. C. 1989. Linking hydrologic and meteorologic components to form an operational flash flood forecast system (abstract). EOS Trans. Am. Geophys. Union 70(15):341.

Smith, J. A., D. J. Seo, M. L. Baeck, and M. D. Hudlow. 1996a. An intercomparison study of NEXRAD precipitation estimates. Water Resour. Res. 32(7):2035-2045.

Smith, J. A., M. L. Baeck, M. Steiner, and A. J. Miller. 1996b. Catastrophic rainfall from an upslope thunderstorm in the Central Appalachians: The Rapidan storm of June 27, 1995. Water Resour. Res. 32(10):3099-3113.

Steiner, M., J. A. Smith, S. J. Burges, and C. V. Alonso. 1997. Use of RADAR for remote monitoring of rainfall rate and rainfall kinetic energy on a variety of scales. Pp. 831-839 in Management of Landscapes Disturbed by Channel Incision. S. S. Y. Wang, E. J. Langendoen, and F. D. Shields, Jr., eds. University, Miss.:University of Mississippi.

Teqnical Services, Inc. 1997. National Electronic Drought Atlas Users Guide. Web site: www.teqservices.com.

Wigmosta, M. S., and S. J. Burges. 1997. An adaptive modeling and monitoring approach to describe the hydrologic behavior of small catchments. J. Hydrol. 202(1-4):48-77.

Wigmosta, M. S., L. W. Vail, and D. P. Lettenmaier. 1994. A distributed hydrology-vegetation model for complex terrain. Water Resour. Res. 30(6):1665-1679.

Woolhiser, D. A. 1996. Search for physically based runoff model—A Hydrologic El Dorado. J. Hydr. Eng. 122(3):122-129.

Woolhiser, D. A., R. E. Smith, and D. C. Goodrich. 1990. KINEROS, A Kinematic Runoff and Erosion Model. Documentation and User Manual, Publication ARS-77. Agricultural Research Service, U.S. Department of Agriculture. Washington, D.C.: U.S. Government Printing Office.

Appendix

Biographical Sketches of Abel Wolman Distinguished Lecturer and Symposium on Hydrologic Sciences

ABEL WOLMAN DISTINGUISHED LECTURER

Thomas Dunne is a professor in the School of Environmental Science and Management and the Department of Geological Sciences and Geography at the University of California, Santa Barbara. He conducts field and theoretical studies of drainage basin, hillslope, and fluvial geomorphology and the application of hydrology and geomorphology to landscape management and hazard analysis. Dr. Dunne obtained a B.A. in geography from Cambridge University in 1964 and a Ph.D. in geography from the Johns Hopkins University in 1969. He conducted research on runoff processes under rainfall and snowmelt while working with the U.S. Department of Agricuture's Agricultural Research Service in Vermont and on hillslope geomorphology while working with the Water Resources Division of the U.S. Geological Survey. Dr. Dunne was elected to the National Academy of Sciences in 1988 for his pioneering work in the fields of geomorphology and hydrology.

SYMPOSIUM CONTRIBUTORS

Diane M. McKnight is a professor in the Department of Civil, Environmental, and Architectural Engineering at the University of Colorado. She received a B.S. in mechanical engineering, an M.S. in civil engineering, and a Ph.D. in environmental engineering from the Massachusetts Institute of Technology. Dr. McKnight was a research scientist at the U.S. Geological Survey's, Water Resources Division, where she studied biogeochemical processes in pristine and mine drainage impacted streams and lakes in the Rocky Mountains. She is a principal investigator for the National Science Foundation's Office of Polar Programs project in Antarctica, where she conducts research on Antarctic lakes. Dr. McKnight is the current president of the American Society of Limnology and Oceanography. She was a coeditor of the book *The Freshwater Imperative: A Research Agenda* (see under references Naiman et al., 1995). Her major research interest is in biogeochemical processes in natural waters.

Eric F. Wood is a professor in the Department of Civil Engineering and Operations Research (Environmental Engineering and Water Resources Program) at Princeton University. His research areas include hydroclimatology with an emphasis on land-atmosphere interactions and the representation of those processes across scales, remote sensing, and hydrologic impacts of climate change. Dr. Wood has received the Horton Award from the American Geophysical Union and is a fellow of the American Geophysical Union and the American Meteorological Society. He is also the chairman of the Hydrology Committee of the American Meteorological Society, a member of the remote sensing committee and of the union fellows committee of the American Geophysical Union, and sits on a number of agency program advisory committees. Dr. Wood received an Sc.D. in civil engineering from the Massachusetts Institute of Technology in 1974.

Fred M. Phillips is a professor of hydrology at the New Mexico Institute of Mining and Technology. He received a B.A. from the University of California, Santa Cruz, in 1976, and an M.S. and a Ph.D. in hydrology from the University of Arizona in 1979 and 1981, respectively. Dr. Phillips' scientific interest is in the area where hydrology, geochemistry, and geology overlap. He received the Clarke Medal from the Geochemical Society in 1988 and was the Birdsall-Dreiss Distinguished Lecturer in Hydrogeology for the Geological Society of America in 1994. He is a member of the American Geophysical Union, the Geological Society of America, the American Quaternary Association, and the Geochemical Society.

Stephen J. Burges is professor of civil engineering at the University of Washington, Seattle. He received a B.Sc. in physics and mathematics and a B.E. in civil engineering from the University of Newcastle, Australia, in 1966. He received an M.S. in 1968 and a Ph.D. in 1970 in civil engineering from Stanford University. Dr. Burges' research interests are in surface water hydrology; urban hydrology; water supply engineering; the application of stochastic methods in water resources engineering; water resources systems, design, analysis, and operation; water resources aspects of civil engineering; and ground water hydrology. He is a fellow member of the American Society of Civil Engineers, the American Association for the Advancement of Science, and the American Geophysical Union. He is the immediate past president of the hydrology section of the American Geophysical Union. Dr. Burges was a member of the Water Science and Technology Board from 1985 to 1989.

SYMPOSIUM DISCUSSANTS

Kaye L. Brubaker is an assistant professor in the Department of Civil Engineering at the University of Maryland. Her work focuses on the strong links between energy and water mass exchanges at the land-atmosphere interface and the implications of these water-energy interactions for regional climate and water resources. She received a B.S. from the University of Maryland in 1989, an S.M. in 1991, and a Ph.D. in 1995 from the Massachusetts Institute of Technology. Dr. Brubaker is a member of the American Geophysical Union, the American Meteorological Society, the American Society of Civil Engineers, and the Soil and Water Conservation Society.

Dara Entekhabi is an associate professor in the Department of Civil Engineering at the Massachusetts Institute of Technology. He received a B.A. in geography in 1983, an M.A. from the Center for Environment, Technology, and Development in 1984; and an M.A. in geography from Clark University in 1987. Dr. Entekhabi received a Ph.D. in civil engineering from the Massachusetts Institute of Technology in 1990. His research interests are in land-atmosphere interactions, remote sensing, physical hydrology, operational hydrology, hydrometeorology, ground water-surface water interaction, and hillslope hydrology. His professional awards include the National Science Foundation's Presidential Young Investigator (1991-1996), the Massachusetts Institute of Technology's Gilbert Winslow Career Development Chair (1994-1996), the Arturo Parisatti International Prize Competition (1994), and the American Geophysical Union's Macelwane Award (1996). Dr. Entekhabi is a member of the American Meteorological Society and the American Society of Civil Engineers and is a fellow of the American Geophysical Union.

David P. Genereux is an assistant professor in the Department of Geology, College of Engineering, Florida International University. He received a B.S. in 1984 from the University of Delaware and an M.S. and a Ph.D. from the Massachusetts Institute of Technology in 1988 and 1991, respectively. Dr. Genereux's research interests include hydrogeology, especially ground water flow and the interaction of ground water with surface water. He is a member of the American Geophysical Union and the American Society of Civil Engineers.

Efi Foufoula-Georgiou is a professor at the St. Anthony Falls Laboratory and the Department of Civil Engineering, University of Minnesota. She received a diploma in 1979 in civil engineering from the National Technical University of

Athens, Greece, and an M.S. and a Ph.D. in environmental engineering in 1982 and 1985, respectively, from the University of Florida. Her major area of interest is hydrology, with an emphasis on understanding and modeling the spatio-temporal organization of hydrologic processes, including precipitation and landforms. Her professional awards include the National Science Foundation's Presidential Young Investigator Award (1989-1994), the National Association of State Universities Commendation for Contributions in Water Resources (1989), and the American Geophysical Union's Excellence in Refereeing (1989). Dr. Foufoula-Georgiou has been an associate editor for *Water Resources Research* and the *Journal of Geophysical Research* and was the main organizer of the Fifth International Conference on Precipitation in 1994. She is the author of 35 refereed publications and an Academic Press-edited volume titled *Wavelets in Geophysics.*